Dryland Degradation:

Causes and Consequences

Edited by Ebbe Poulsen
and Jonas Lawesson

AARHUS UNIVERSITY PRESS

AARHUS UNIVERSITY PRESS
Building 170, Aarhus University
DK-8000 Aarhus C, Denmark

Introduction

This volume contains abstracts and papers presented at the second Danish Sahel Workshop, held at Sandbjerg Manor in January, 1990.

The aim of the Danish Sahel Workshops is to gather Danish scientists, administrators and consultants currently working with the Sahel and related areas in East and Southern Africa. In order to widen the perspective and experience represented, guests from other Nordic countries and Great Britain were invited.

The deliberations of the workshop reflected the wide range of Danish Sahel activities, with papers covering natural and social sciences, such as improved utilization of tree legumes, degradation and desertification, afforestation techniques and problems, ecological monitoring and remote sensing techniques, animal counting in Senegal, the political economy of agricultural development in Burkina Faso, research and development in the Red Sea area, the economic potential of the semi-arid areas, as well as accounts of the research sponsored by DANIDA, integrated projects in Mali and Senegal and the Danish activities with control of parasitic diseases in Africa.

Most of the costs of the workshop were generously covered by The Danish Research Council for Development Studies (DANIDA). The present publication has been financed through a grant from the Aarhus University Research Foundation. Ms. Anni Sloth assisted with the figures and Ms. Jette Nieheleather Hulcahy with linguistic corrections.

Aarhus, February 1991

Jonas Lawesson, Ebbe Poulsen

Contents

Desertification and Land Degradation in Perspective

Lennart Olsson

Abstract

This article critically reviews the international debate on desertification, its status, its rate of growth as well as its causes and consequences. An approach which applies systems analysis of the entire socio-ecosystem to the study of desertification and land degradation is discussed, using examples from Western Sudan.

Do we have enough knowledge to enable us to push back deserts? Professor Mohammed Kassas, Senior Advisor to UNEP Executive says: Yes." (UNEP 1979 p. 23)

Significant illustration of UNEP's conception of desertification.

1. Introduction

Desertification, or land degradation as I prefer to call it, has been described in many catchy ways :

We now must stop the advance of the desert... in Mali the Sahara has been drawn 350 kilometers south by desertification over the past 20 years. (The President of the World Bank (Forse 1989))

The Sahara Desert continues to creep southward, claiming an area the size of New York State every decade. (Smith, 1986)

It has been estimated that 650,000 square kilometers of the Sudan had been desertified over the last 50 years and that the front line has been advancing at a rate of 90-100 kilometers annually during the last 19 years (Suliman 1988).

On the 14th of March, 1986, that Vice-President Bush was being urged to give aid to the Sudan because "desertification was advancing at 9 km per annum" (Warren & Agnew 19980)

Currently, 35 per cent of the world's land surface is at risk... each year, 21 million hectares is reduced to near or complete uselessness. (UNEP 1984)

The best recent example of desert encroachment occurred with the Sahelian drought between 1968 and 1974, when climatically and ecologically it was as if the Sahara had extended its limit southward by 5 degrees of latitude. (Tyson 1980).

Desertification ... is probably the greatest single environmental threat to the future well-being of the earth (Högel 1979).

These are a few of the many sweeping statements on desertification recently published in popular as well as respected scientific journals. Desertification has been a buzzword in the development debate since the UN Conference on Desertification in 1977. The vaguely and ambiguously defined process of desertification was made the scapegoat of the food crisis in Africa as well as a number of other problems. In the different attempts to define this extremely complex process, causes and effects were often mixed up. Perhaps one of the gravest mixed-up factors is poverty. A number of articles claim poverty being the effect of desertification. I would like to stress firmly that poverty is the ultimate cause of land degradation. People are forced, by lack of food and other necessities, to exploit the natural environment.

Because of the lack of proper definitions, the whole concept of desertification became much misused. The main body responsible for the co-ordination of research into and counteracting desertification, the Desertification Control Programme within UNEP, defines desertification as:

the diminution or destruction of the biological potential of the land, and can lead ultimately to desert-like conditions (Mabbutt 1985).

The main ambiguity in this definition is "desert-like". Most of the dry-lands (arid, semi-arid and humid areas) can give a "desert-like" impression on any western traveler during the dry season (the season for many expert missions on this theme) or during periods of drought. One of the most influential and misleading studies was written by Lamprey (1975). He compared the position of the desert boundary in western Sudan at two different points in time. The first was the desert boundary marked on a vegetation map in 1958 (Harrison & Jackson), and the second was the result of a light aircraft reconnaissance carried out

6

in 1975. Lamprey found that the desert edge had advanced 90-100 km southward in 17 years.

1. 1975 was just after the very severe Sahelian drought (1968-74). The report failed to distinguish between the temporary effects of the drought and a secular trend. The Harrison & Jackson vegetation map, on the other hand, was compiled during the 1950s, a period of very favorable rainfall throughout the Sahelian zone.

2. The two studies used different definitions when drawing the desert boundary. Harrison & Jackson used the 75 mm isohyet (Harrison & Jackson 1958), drawn from a very scarce and inaccurate network of rainfall stations, while Lamprey used field and areal observations of vegetation conditions.

The different quantitative statements of a certain number of hectares being desertified or lost annually, emanate probably from the estimates carried out in 1977 and in 1984. However, no systematic world survey of the status and trend of desertification was carried out for the 1977 UN Conference on Desertification. Such estimates were only by-products of the mapping of world desertification, which was not based on any primary research. The second estimate, carried out in 1984, was based on two sources; a questionnaire administered to 91 affected countries and to 10 donor countries, and regional assessments conducted by the UN Regional Commissions. In an evaluation of the two estimates, Mabbutt (1985) points out the weaknesses and poor reliability of the data. However, he then goes into a detailed analysis of the estimates. Since then, these extremely rough figures have been quoted and manipulated by other authors. Some of the more ridiculous measures of desertification rate, use number of football grounds per time unit. These articles, aimed at popularizing the severeness of land degradation, contributed to the false picture of an advancing desert. This view of desertification gave birth to a number of more or less useless attempts to stop the "march of the desert", like green belts to halt sand movement or sand dune fixation.

The grave exaggerations concerning the area subject to desertification as well as the rate of increase had a number of negative effects on the dryland communities. I would like to stress that many exaggerations were made with a good purpose - to increase the flow of international aid to the drought and famine stricken communities. However, this policy backfired in the mid and late 1980s, when the World Bank and other major donor organizations more or less cut off the support to the Sahelian zone. The reason for this, was simply that those countries were destined to become deserts in the near future. A big change in this policy took place in 1989, to a large extent after the completion of a report by Ridley Nelson of the World Bank (Nelson 1988). The report concluded that the problem of desertification had been gravely exaggerated and that the evidence of desertification was extremely scanty. Another important conclusion was that "the availability of profitable technologies to combat the problem has been overestimated" (Nelson 1988). The report recommends that the World Bank increase its support to dryland communities, and emphasizes the need for profound research in order to reach a deeper understanding of the causes and effects of desertification. Another report that expressed similar ideas was written by Warren & Agnew (1989). Furthermore, the criticism surrounding the popular view of desertification was reviewed in a short paper by Forse (1989) which had a significant impact on the international debate.

Another ambiguity in the desertification debate has been the question of whether the process is reversible or not. There are examples of both. Some of the salinized lands of Babylonia are still unproductive, while the lands of the "dust bowl" disaster in USA have been successfully restored and recovered as productive range land in just a few years (Rapp 1987). However, the discussion on the reversibility of desertification leads to the concepts of resilience and stability within the socio-ecosystem.

Stability of a system can be defined as its ability to remain the same while external conditions change (Noy-Meir & Walker 1986).

Resilience was defined by Holling in the following way:

Resilience determines the persistence of relationships within a system and is a measure of the ability of these systems to absorb changes ... it is the property of the system and the persistence or probability of extinction is the result. (Holling 1973)

Another important concept in this context is sustainability. A certain kind of land use is sustainable if it can continue indefinitely. We can therefore say that resilience of a system is the property that determines the sustainability of its use. "Land degradation, very simply, is loss of resilience" (Warren & Agnew 1988).

A factor that is of special importance in dryland systems is spatial diversity, mainly caused by the erratic nature of rainfall. The high degree of resilience of the traditional nomadic socio-ecosystem, was based on the ability to move over large areas every season to make use of the large spatial variation which is a characteristic feature of all arid and semi-arid environments. With restricted spatial mobility, caused by e.g. conflict over landownership with the sedentary people, the traditional pastoral system lost much of its resilience. In the case of sedentary farmers, seasonal migration of the male population to other regions has always been an important factor increasing the resilience of the sedentary farming socio-ecosystem.

During periods of drought, like the Sahelian drought 1968-74 and the very severe 1984-85 drought of NE Africa, the socio-ecosystems collapsed in many cases - they were not resilient enough to cope with the extreme variations in rainfall. The systems had either lost resilience, or the variation in rainfall was larger than previously experienced.

2. Research on Land Degradation

During the 1970s, much of the research was devoted to monitoring the process of desertification. Due to the lack of proper definitions and the lack of knowledge regarding the many interrelated factors, the research became focused on indicators of desertification. The many indicators proposed (Reining 1978, Dregne 1983) were seldom related to any conceptual model of the socio-ecosystem under study. Attempts to build deterministic models based on these indicators (Hellden 1981) failed. A major problem of this "indicator approach" is that virtually all indicators are, to varying extent, dependent upon each other. However, in practical primary research it may be necessary to use indicators in order to obtain tangible information. There is often conflict between what is "doable" for the individual researcher and what is relevant in a wider context.

2.1 The Importance of a Holistic Approach

In the holistic approach, a conceptual model of the system under study is necessary. In a conceptual model, we concentrate on the relationships between the system components rather than the value or function of each component. In this conceptual model of the socio-ecosystem we must try to define the factors adding to its resilience. In Agnew (1984) a possible conceptual model of the Sahel food production system was presented. In addition to the model presented by Agnew, this approach merchants as an extra component. Although the merchant community does not produce any food, it is an extremely important group, sometimes possessing total control over markets and prices. A well

9

functioning market system with a sound pricing strategy contributes significantly to the system's resilience - and vice versa.

The conceptual can be broken up into a number of possible sub-systems, and factors adding to each sub-system's resilience.

Table 1. Possible subsystems of a dryland socio-ecosystem, and the factors adding to its resilience.

Vegetational system	- drought resistant species
	- drought evading species
	- fire resistant species
Soil system	- sandy soils - high rainfall acceptance
Farming system	- drought resistant crops
	- drought evading crops
	- mixture of crops
	- fallow periods
	- mixed agriculture and animal husbandry
	- mixed agriculture and gum tapping
Economic system	- cash crop sale
	- capital accumulation
	- availability of credits
	- food storage
	- emergency food aid
Social system	- migration possibilities
	- network of relatives

With an appropriate conceptual model of the system, it may be possible to select critical indicators which can become topics for primary research.

3. Is Land Degradation (Desertification) a Myth?

As claimed earlier in this article, the popular view of desertification as a moving desert, burying fertile land and human settlements, is false. If we stick to the UNEP definition of desertification, what is the evidence of diminution or destruction of the biological potential? Apart from e.g. formation of salt crust, exposure of infertile horizons, and severe erosion there is little evidence that the biological potential of the land is being reduced. What we can see, is a very clear variation of the biological production with rainfall.

I will give a few examples at different scales below.

1. Continental scale:
Using data from the NOAA satellite, Dregne & Tucker (1988) found a northward movement of vegetation boundaries along the southern edge of the Sahara of 150 to 200 km from 1984 to 1985. 1984 was one of the driest years in many decades, while 1985 was unusually wet. The study is only a comparison of two years, but it is at least a strong indication of the correlation between rainfall and vegetation

2. Regional scale:
There is a very high correlation between crop yield (i.e. biological productivity of the land) and rainfall. Olsson (1983 & 1985) found that roughly 70% of the variation in millet yield in the Northern Kordofan province of the Sudan could be explained by rainfall parameters (primarily, number of days with at least 1 mm rainfall).

4. A Second Equilibrium Hypothesis

A system that has been subject to change of a given magnitude can either recover towards its original state, or it can move towards some other state. I would like to put forward the following hypothesis, valid for the sandy regions of the Northern Kordofan and Northern Dafur provinces in the Sudan:

The biological potential was probably higher many years ago, when the traditional long fallow periods were maintained in the farming systems. During the fallow periods, averaging at 17 years according to Hammer-Digernes (1977), the nutrient status of the soil improved. The land was then used for cultivation, of millet or sesame for example. After only 3 to 5 years, cultivation was abandoned because of a dramatic drop in yield. This drop in yield was mainly caused by the depletion of nutrients, and the infestation of pests, Striga (or "witchweed") for example. This means that the ecosystem changed into another state after only a few years of intensive cultivation. Fallow periods in this region have now been drastically shortened, mainly due to increased population density. There is now no, or at least not enough, time to recharge the nutrient status of the soil between the cultivation periods, resulting in very poor soils. Soil samples taken systematically over a large area in the Northern Kordofan Province, show extremely low nutrient contents in both cultivated and fallow fields (Olsson 1985). There was virtually no difference between the nutrient content of the top soil and that of the subsoil, which is generally expected.

The system has moved into a second equilibrium, where the soils are equally poor throughout the horizon. The biological potential of this land is low, but further erosion and other degrading processes do not effect the biological potential any longer, provided that the soil is deep enough. The vegetational system has now become almost entirely dependent on rainfall. The system lost resilience when it moved from a farming system with long fallow periods to a system with short fallow periods or even continuous cultivation. But that transition took place more than two decades ago. To re-establish the old farming system with long fallow periods is of course highly desirable but hardly possible without a significant structural change to the regional and national economic system. Meanwhile it is important to find other means of strengthening the resilience of the socio-ecosystem.

5. Concluding Remark

Poverty and food security are fatal problems in many African drylands. Desertification has, during the last two decades, been blamed as one of the most important roots of these problems. Due to inappropriate definition, the concept of desertification has been misused and exaggerated. This has obscured some of the most crucial questions and prevented efficient action. Research has often been fragmented, superficial and narrow. A systems analysis approach, where a system is considered in its widest perspective, could assist us in addressing the crucial question: to increase the resilience of dryland socio-ecosystems in order to attain the sustainable use of nature and food security.

References

Agnew, C.T. 1984: Checkland's soft systems approach - a methodology for geographers? Area vol. 16, no. 2: 167-174

Dregne, H.E. 1983: Desertification of Arid lands. Harwood Academic Publishers, New York, 242 p.

Dregne, H.E. & Tucker, C.J. 1988: Desert encroachment. Desertification Control Bulletin 16: 16-20

FAO, UNESCO & WMO, 1977: World map of Desertification at a scale of 1:25,000,000, Explanatory Note. UNCOD, A/CONF 74/31, New York.

Forse, B. 1989: The myth of the marching desert. New Scientist, 4 February 1989: 31-32

Gregory, K.J. & Walling, D.E. (eds) 1987: Human Activity and Environmental Processes. John Wiley, New York.

Hammer-Digernes, T. 1977: Wood for fuel - Energy Crisis Implying Desertification. Geografisk Institutt, Universitetet i Bergen, Norway.

Harrison, M.N. & Jackson, J.K. 1958: Ecological Classification of the vegetation of Sudan. Forests Bulletin No. 2, Khartoum, and the Vegetation map of the Sudan. Sudan Survey Department, Khartoum.

Hellden, U. 1981: Satellite data for regional studies of desertification and its control - approaches to rehabilitation of degraded ecosystems in Africa. Lund University, Dept. of Physical Geography Paper no. 50.

Högel, J. 1979: Introduction to Desertification Control Bulletin, vol. 2 no. 1.

Holling, C.S. 1973: Resilience and stability of ecological systems. Annual Review of Ecology and Systematics 4: 1-23.

Joss, P.J., Lynch, P.W. & Williams, O.B. (eds) 1986: Rangelands: a Resource under siege. Proceedings of the 2nd Int. Rangeland Congress. Australian Academy of Science, Canberra, 630 p.

Lamprey, H.F. 1975: Report on the Desert Encroachment Reconnaissance in Northern Sudan: 21 October to 10 November 1975. Published in Desertification Control Bulletin 17.

Mabbutt, J.A. 1985: Desertification of the world's rangelands. Desertification Control Bulletin 12, pp. 1-11

Mabbutt, J.A. & Berkowicz, S.M. 1980 (eds): The threatened Drylands, Proc. Pre-Congress Symposium C19, International Geographical Congress, Japan

Nelson, R. 1988: Dryland Management: The "Desertification" Problem. Environmental Dept. Working Paper no. 8, The World Bank, Washington.

Noy-Meir, I. & Walker B. H. 1986: Stability and resilience in rangelands, in Joss et al. 1986, pp. 21-25.

Olsson, L. 1983: Desertification or Climate? Investigation regarding the relationship between land degradation and climate in the central sudan. Lund Studies in Geography, Ser. A, no. 60. Lund.

Olsson, L. 1985: An Integrated Study of Desertification. Lund Studies in Geography, Ser. C, no. 13. Lund.

Rapp, A. 1987: Desertification. Chapter 17 in Gregory & Walling 1987.

Reining, P. 1978: Handbook on Desertification Indicators. American Association for the Advancement of Science, Washington D.C.

Smith, S.E. 1986: Drought and water management: the Egyptian response. Journal of Soil and Water Conservation 41: 297-300.

Suliman, M.M. 1988: Dynamics of Range Plants and Desertification Monitoring in the Sudan. Desertification Control Bulletin 16: 27-31.

Tyson, P.D. 1980: Climate and Desertification in Southern Africa, in Mabbutt & Berkowicz 1980, pp. 33-44.

UNEP 1979: Desertification Control Bulletin, vol. 2, no. 1

UNEP 1984: Activities of the United Nations Environment Programme in the combat against desertification. A report prepared by the Desertification Branch of UNEP. Nairobi.

Warren, A. & Agnew, C. 1988: An Assessment of Desertification and Land Degradation in Arid and Semi-arid Areas. Paper no. 2, International Inst. for Environment and Development.

Vegetation Changes in Senegal.
A Photographic Perspective

Jonas Lawesson

Trends in the vegetation succession of Sahel are little known. In rainy years, the greenness of the vegetation is anticipated as the first result of "fighting the desert", whereas a number of dry years are often claimed to be an indication of "desertification" - a quickly progressing, irreversible process in the marginal zone of Sahara in Africa. Besides ambiguous definitions of "desertification" (Verstraete, 1986) floristic or vegetational indications of such "desertification" are hard to detect in Senegal and Sudan at present, as borne out in some recent studies (Lawesson, 1990; Olsson, this volume) and even more clear when photographic evidence is included. It seems interesting in this context to compare old photographs from Sahel with present day photographs, taken by the author during the Aarhus University Sahel Project in Senegal 1988-90. The old photographs have been found in the monumental work by Trochain (1940) on the vegetation of Senegal. The author's photographs generally attempt to match the same areas as the old ones.

With this photographic comparison, it may be possible to detect the similarities *and* changes in the vegetation, not in terms of floristic composition, but in structure and aspect. Certainly changes have occurred, as the result of natural and man-made processes, and there is a general consensus that some kind of degradation is occurring. However, the nature of this degradation and the ability of the ecosystem to counteract it is much discussed.

The selection of photographs, here presented, may eventually make us consider the validity of many of the alarming statements arriving to us from scientists and journalists alike, and remember to utilize any historical sources, if they exist.

Fig. 1.1. "Pseudoclimax with *Acacia Raddiana* (= *tortilis*), between Louga and Linguére, February 23, 1934" (Trochain, 1940)

Fig. 1.2. Wooded grassland with *Acacia tortilis* and *Acacia senegal*, near Linguére, 1989.

Fig. 2.1. "Sahelian mare, Niakka, close to Barkedji, with *Acacia scorpioides var. pubescens* (= *A. nilotica*). November 25, 1930" (Trochain, 1940)

Fig. 2.2. Sahelian mare close to Linguére, with *Acacia nilotica, Acacia tortilis, Balanites aegyptiaca* and *Celtis integrifolia*, 1989.

16

Fig. 3.1. "Sahelian savannah with *Combretum glutinosum* as dominant, near Raneirou, July 8, 1934" (Trochain, 1940)

Fig. 3.2. Wooded grasssland with *Combretum glutinosum, C. nigricans* and others as dominant species, Diennédié, 1988.

Fig. 4.1. "Cover of *Pterocarpus lucens,* at thin soil, in Ferlo, March 13, 1934" (Trochain, 1940)

Fig. 4.2. Wooded grassland with *Pterocarpus lucens, Commiphora africana*, East of Linguére, 1989.

Fig. 5.1. "Forest savannah between Goudiry and Tambacounda, June 18, 1934" (Trochain, 1940)

Fig. 5.2. Woodland between Goudiry and Tambacounda, 1989.

19

Fig. 6.1. "Soudano-Sahelian savannah North of Goudiry, June 11, 1934" (Trochain, 1940)

Fig. 6.2. Woodland-grassland North of Tambacounda, 1989.

References

Lawesson, J.E. 1990: Sahelian woody vegetation in Senegal. Vegetatio 86: 161-174.

Lawesson, J.E., Freudenberger, M.S. & Tybirk, K. (in press): Ecology and economy of the Gum Arabic Tree (Acacia seneggal) in Northern Senegal.

Olsson, L. 1991: Desertification and land degradation in perspective. This volume.

Trochain, J.L. 1940: Contribution a l'étude de la vegetation de Senegal. Memoirs de l'IFAN: 2, IFAN and Larose, Dakar and Paris.

Verstraete, M.V. 1986: Defining desertification: A review. Climatic Change 9: 5-18.

Planting trees in Sahel
- doomed to failure?

Knud Tybirk

Introduction

Limitations of the natural regeneration of trees as well as the traditional ways of living in Sahel have to be incorporated in planning aimed
at reaching the long term goals of attempts to reestablish tree growth.

Tree planting in dry areas is often expensive and the long-term
results are at best sparse and, unfortunately, often even lacking, despite
a rather high financial input. However, at present the development
agencies (both international, national and NGO's) are motivated to ini-
tiate large-scale tree planting programmes in Sahel. The key question
is, whether this new green policy merely benefits the image of the
developmental organizations without positive long-term results?

Traditional Tree Planting

Tree planting has traditionally been carried out in more or less the
same way, whether it has been a project financed by an international
donor or a national program. Seed sources have to be localized followed by collection, cleaning and storage (often using pesticides against
seed predators) of seeds. To obtain fast and homogeneous germination,
the seeds often have to be pretreated by either boiling, acid soaking or
manual nicking or scarifying of the seed coat. Seeds are sown in plastic
bags sheltered from the sun and kept moist by daily watering (Doran
et al. 1983).

After germination and several months of careful nursing, the seedlings are ready for transplanting. This phase is critical in the life of the
young plants and many seedlings will die or suffer severely from the
dramatic change (Gosseye 1980). If the plant roots reach deeper and
more moist layers in the soil during the first rainy season, then the
plants have a chance of surviving the first dry season.

New Priority Species

In the late seventies an important change occurred in FAO and later in the national and international organizations concerned with forestry. FAO changed some of their principles for selecting tree species for planting in dry areas. "Multipurpose tree species" was the name for tree species with multiple uses (fodder, firewood, medicine etc.) locally adapted and known by local people (Graudal 1989). Such characteristics were given high priority instead of homogeneity, growth rate and industrial uses, which had been key words in proper forestry until 10-15 years ago.

Many trees of the Legume family (Leguminosae) are locally known and appreciated for their multiple uses and drought tolerance. Many of these became priority species used by FAO in the drier areas of Africa (Graudal 1989). Especially many species of the widespread genera Acacia and Prosopis were promoted and some biological and silvicultural information was gathered (Brenan 1983, Doran et al. 1983, Johnson 1983, Southgate 1983). Some of these species have been used locally for decades for tree planting, the results being rather variable (Greaves 1984a, b, Giffard 1964). Experiments for germination and transplanting techniques have been repeated several times, but the results have never been related to the ecological conditions, and rarely related to the way of life of people in the areas used for planting.

Research

Through studies of related literature and fieldwork in Kenya and Senegal, the author of this article has tried to review the existing knowledge about the natural and artificial modes of legume tree regeneration and to relate this to the ecological and human conditions found in Sahel.

The conclusion of this work (Tybirk, in preparation) is that the traditional method of planting legume species will rarely be succesful on a long-term basis. The present study has included 35 common species of legume trees and shrubs found in Sahel, all under similar conditions (Maydell 1986). The natural regeneration pattern of these species has been clarified and this basic knowledge about natural bottlenecks in their lifecycle provide useful information for future management plans of woody resources in Sahel. The regeneration pattern of the trees is closely related to the human management of the areas in which they are cultivated, but the most important factor is, above all, the variable climate, which still lies beyond human control. Therefore, humans, trees, and climate are integrated parts of a fragile ecological system, in which the human needs and management are crucial for the future. Planting trees and expecting a reestablishment of the former ecological balance is not possible without some kind of adaptation occurring in the living habits of the people in the areas concerned. Furthermore, the natural regeneration of the trees has to be considered to be able to ensure natural regrowth of the vegetation and thus a stable supply of forestry products.

Let me here shortly summarize the four main bottlenecks (dispersal, seed predation, germination and seedling growth) influencing the natural regeneration of legume trees and shrubs in Sahel.

Bottlenecks

Seed dispersal of the trees in question primarily takes place in two ways. About half of the species are adapted to wind dispersal, while the rest are dispersed by the consumption of the pods by cattle and some of the seeds passing through undamaged (Coe & Coe 1987). The digestive juices of the animals often enhance germination by softening the seed coat in the same way as the mentioned acid treatments in the nursery. The digestive juice may even reduce the predation of seed boring beetles. The feces of the cattle can also have a positive effect on seed germination and establishment of seedlings. However, the combined effects of chewing, beetle attack, digestive juice etc. softens the seed coat of the weakest seeds, which are digested and serve as an extra reward for the cattle for acting as disperser of the surviving seeds (Schmidt 1988).

The second natural bottleneck is seed predation. In nature, all the included species are attacked by seed boring bruchid beetles (Coleoptera: Bruchidae). Often the seed crop of the year is reduced and particularly the seeds in stores are destroyed by these beetles. In a balanced natural system the beetles and their associated parasites act as natural regulators of the woody legumes (Ernst et al. 1989). However, only little information is available in general about the taxonomy, distribution, host preference etc. of bruchids (Johnson 1981, 1983, Southgate 1983) and their influence on natural systems, but they are quite important in traditional forestry when collecting and storing seeds. Often pesticides have been used to reduce attacks on stored seeds, but other methods have proved to be useful, e.g., airtight or carbon dioxide-filled storage containers (Johnson 1983).

Germination is the next critical phase for legume seeds. The seeds

24

of woody legumes in Sahel are adapted to the variable and harsh climatic conditions. The hard seed coat protects the seed against desiccation, against digestive juice of dispersers, against fire, some predators etc, and the seed coat is therefore the most important factor for regulation of germination (Cavanagh 1980, Schmidt 1988). It is crucial to germinate on the right time and place for the survival of the seedlings during the first year. At the same time it is important not to put all the eggs in one basket, i.e. all the seeds in one crop should not germinate in the same year, as all of the seedlings would die if a dry period were to follow the germination period. Because of their hard seed coats, some of the seeds will presumably survive, lying dormant for many years in the soil, awaiting the right moment when good conditions for germination will be present. Many questions concerning the mechanisms for water penetration of the seed coat and germination initiation are unclear, hence this field awaits more research in the future.

The juicy and nutritious seedling is an attractive food resource for many animals (ungulates, rodents, insect larvae, grasshoppers etc.), and the first months of growth are the most vulnerable part of the life-cycle of the trees. Domesticated animals in Sahel, especially sheep and goats, are often responsible for the low percentage of seedling survival. Some species can survive one browsing of the leaves, but repeated browsing during the growth season will kill the plant. Fires can also kill or reduce the growth of seedlings, but if it happens in the middle of the dry season, when growth is lacking, the plant will often be able to resprout during the following rainy season (Giffard 1964, 1966, 1971).

The most important factor in this part of the life-cycle is a good rainy season, giving the seedling the opportunity to develop deep roots to reach lower and more humid soil layers before the following dry season. Often several consecutive good rainy seasons are needed, but after 2-5 years the plants may have attained a height of 2-3 meters and will be out of reach of goats and fires. From then on they have a good probability of surviving and developing into mature trees.

"Integrated Natural Management"

The phases in the life-cycle of woody legumes in Sahel just described show how small the chances are for a seed to survive dispersal without being digested, to germinate under good conditions, to grow up without being trampled, eaten, burned, or suffering from drought. These natural conditions are important to keep in mind, when planning to reestablish the woody vegetation in a particular area.

Theoretically it is possible to establish vegetation by traditional tree planting, but often the results after five years are disappointing compared to the input made. Here, a proposal is presented whereby a much lesser investment is employed to reach the same or better results by integrating the natural conditions in agreement and cooperation with the local people.

The following is a theoretical example which may serve as a framework for ideas to be used in future planning, which may be added to and detracted from.

An Example

An area in any Sahelian country has been depleted of most of its former natural savanna vegetation, eplaced by grass layer and dispersed trees, mostly acacias. The National Forestry Department has made an agreement with the local semi-nomadic population and an external donor to reestablish a "natural" vegetation, to the benefit of the local people using the trees and for the natural environment.

On the basis of the nature and the people in the area concerned the project seeks to reestablish the mixed woody vegetation of, let us say, a mixture of Acacia nilotica, A. tortilis, A. seyal, Bauhinia rufescens and Dicrostachys cinerea. Local herdsmen are paid to collect whole pods of these species at the end of the rainy season, and the pods are kept in airtight or carbon dioxide-filled containers (to reduce attacks by bruchid beetles) by the Forestry Department until the end of the dry season.

Late in the dry season, an agreement is made with the local herdsmen to rent their herds of cattle, goats and sheep for a fortnight or so. In selected good localities small areas - of perhaps half a hectare each - are fenced. These areas have to be small and distributed over a large area, to interfere as little as possible with the traditional pastureland of the herdsmen, who themselves have to be integrated into the decision-making process and thereby motivated by the short and longer term benefits to their own situation. The rented animals are fenced in, fed with the pods collected earlier and left defecating and trampling in the enclosures for a week or so (fed and watered properly of course). In this way the undigested seeds have been pretreated and spread in a newly fertilized and disturbed soil where they have excellent conditions for germination and growth when the rains come.

The cattle are released, but the fencing is kept in place during the next 2-3 years to protect the seedlings from trampling and browsing. Then you just have to await sufficient rainfall and hope that some of the seeds will develop into mature plants. Ideally, after some years a small grove is established, which after a decade or so will be capable of spreading over larger areas in good years if - and only if - the local people can see the benefits of the project and try to assist the "natural" regeneration.

This theoretical example will of course only be successful in a few cases, but it is my firm belief that this kind of project can be just as successful as traditional tree planting projects. The input here is much less and the work can be done by local people. If the trial is unsuccessful, it can simply be repeated the following year. More of the local conditions are taken into account, and it is easier to motivate local herdsmen to cooperate when the project acts closely in accordance with their normal way of living. To put it another way: this kind of forestry will not violate either the lifestyle of the people nor the natural process of woody regeneration.

Patience

The dry periods of the last two decades have placed focus on the concept of desertification. However, it has never been proved that this phenomenon not only reflects the natural climatic variations in Sahel (Olsson, this volume) but is also, somehow, made worse by growing pressure from the population. The most important conclusion about the regrowth of trees in Sahel is that the trees are well adapted to these natural climatic variations. In spite of the many odds against natural regeneration it still happens in some favorable areas and years. The natural resilience of the woody legumes in Sahel indeed makes regrowth possible in a decade with above average precipitation.

The future of the woody vegetation in Sahel is thus dependent on the integration of the adaptive processes of trees to the living practices of people under the unstable climatic conditions. This will demand the practice of greater patience by the national and international donors than that necessitated by the 3-5 year tree planting projects hitherto planned to ensure a stable supply of forestry products in the future.

References

Brenan, J.P.M. 1983: Manual taxonomy of Acacia species. FAO, Rome.

Cavanagh, A.K. 1980: A review of some aspects of germination of acacias. Proc. Roy. Soc. Vic. 91: 161 - 180.

Coe, M. & C. 1987: Large herbivores, acacia trees and bruchid beetles. S. Afr. J. Sci. 83: 624 - 635.

Doran, J.C., Turnbull, J.W., Boland, D.J., & Gunn, B.V. 1983: Handbook on seeds of dry-zone acacias. FAO, Rome.

Ernst, W.H.O., Tolsma, D.J. & Decelle, J.E. 1989: Predation of seeds of Acacia tortilis by insects. Oikos 54: 294 - 300.

Giffard, P.L. 1964: Les possibilités de reboisement en Acacia albida au Sénégal. Bois et Forêts des Tropiques 95: 21 - 33.

Giffard, P.L. 1966: Les Gommiers: Acacia senegal Willd., Acacia laeta R. Br. Bois et Forêts des Tropiques 105: 21 - 32.

Giffard, P.L. 1971: Recherches complémentaires sur Acacia albida (Del.). Bois et Forêts des Tropiques 135: 3 - 20.

Gosseye, P. 1980: Introduction of browse plants in the Sahelo-Sudanian zone. In Le Houârou, H.N. (Eds), Browse in Africa. ILCA, Addis Abeba, pp. 393-397.

Graudal, L. 1989: Development and evaluation of genetic resources of multipurpose woody species for dry areas with special emphasis on northern Sudano-Sahelian Africa, in: Tybirk, K., Lawesson, J.E.L. & Nielsen, I. (eds.), Sahel Workshop 1989 (AAU reports 19, Botanical Institute, University of Aarhus) pp. 56 - 61.

Greaves, A. 1984a: Acacia nilotica. Commonwealth Agricultural Bureau, Slough.

Greaves, A. 1984b: Acacia tortilis. Commonwealth Agricultural Bureau, Slough.

Johnson, C.D. 1981: Seed beetle host specifity and the systematics of the Leguminosae, in: Polhill, R.M. & Raven, P. (eds), Advances in Legume Systematics. Royal Bot. Gard. Kew, pp. 995 - 1027.

Johnson, C.D. 1983: Handbook on seed insects of Prosopis species. FAO, Rome.

von Maydell, H.J. 1986: Trees and shrubs of the Sahel. Their characteristics and uses. Schriftenreihe der GTZ. GTZ Verlagsgesellschaft, Rossdorf.

Olsson, L. 1991: Desertification and land degradation in perspective. This volume.

Schmidt, L.H. 1988: A study of natural regeneration in transitional lowland rainforest and dry bushland in Kenya. (Thesis, Botanical Institute, University of Aarhus).

Southgate, B.J. 1983: Handbook on seed insects of Acacia species. FAO, Rome.

Tybirk, K. (in prep.): Handbook on Regeneration of Sahelian Woody Legumes (English/French).

Vegetation Changes in Arid and Semi-arid Africa

Christina Skarpe

Introduction

Vegetation is consistently changing. Indications on changes of different scale in the arid and semi-arid zones of Africa can be obtained from (sub)fossil and archaeological evidence, oral tradition, written records, living memory and direct monitoring and research. Reasons for change can be geological or evolutionary processes, climatic change, human land use or interaction between many factors. As the time-span under consideration becomes shorter and more recent, the significance of direct and indirect impact by man increases.

Prehistoric rock paintings in Sahara illustrate some aspects of a transition from lush savanna with elephants and buffalos to arid conditions with ostriches and camel riders during some 10,000 years of climatic change and human influence. The process is presently repeated in the corresponding ecological zones of Southern Africa during a few centuries, and with little evidence of climatic change.

For resource development and land use strategies, the main interest may be on changes during the last decades or, at most, century. In that time perspective, in most cases, human impact is considered the prime agent. (There may be too little data to tell the nature of the supposed decrease in rainfall in the Sahel since the mid 1960's).

The work on vegetation changes in arid and semi-arid regions has to a large extent concentrated on the so-called "desertification" problem, and has included much mapping, often using satellite imaging or other kinds of remote sensing, of the border between vegetation and bare soil. Less is known about the vegetation as such and its dynamics, and still less about the mechanisms governing its changes. It is symptomatic that last year's comprehensive report on the IUCN Sahel programme lacked a chapter on range condition and development, and stated that this omission was because the information simply does not exist.

Regarding the importance of livestock for human subsistence in all dry parts of Africa, it is essential to fill this gap. In order to manage and use the grazing resources well, local vegetation descriptions and range monitoring as well as an improved understanding of the mechanisms governing vegetation dynamics are important.

Savanna Eco-Systems

Savannas or savanna-like vegetation types cover more than half of Africa. Most of them are man-made, anthropogenic savannas, and also the natural, dry ones are usually considerably modified by human impact. Savannas are primarily found in areas with summer rain and pronounced dry season(s). The mean annual rainfall is often between 2,000 and 200 mm. Four main determinants in savanna eco-systems are available: soil water, available nutrients, fire and herbivore. Man directly interferes with the frequency and time of fires and kind and degree of herbivore. In this way he indirectly influences also the availability of and competition for water and nutrients. Man also cuts trees and cultivates fields.

A general model of savanna eco-systems is presented by Walker & Noy-Meir (1982). It describes a savanna as a two-component and two-layer system. It has a continuous herbaceous layer, mainly grasses, and a discontinuous layer of woody species, trees and/or scrubs. The main limiting factor is water. Both grasses and woody species have access to water in the surface soil, but grasses can outcompete woody species for water in that zone. The water penetrating to deeper soil layers is exclusively accessible for woody species. Of course, nature is more complex than this model, but in a generalized way it seems to well describe important competition patterns.

Other factors being constant, the amount of woody growth will, according to the model, increase if grasses decrease, while grasses are less affected by changes in the woody component. The isoclines for woody vegetation and grasses have one stable point of equilibrium, implying a natural balance between the two elements.

Changes in the Ratio Between Grasses and Woody Vegetation

Throughout the world there is much evidence for changes in the ratio between the woody and the grassy component in dry rangelands. In most cases it involves an increase in scrubs at the expense of the more open grass dominated savanna. The resulting thicket vegetation is in most cases of little use to man, offering no grazing, usually low quality browse and too weak stems to provide good wood. The change in vegetation is usually attributed to human disturbance, primarily to over-grazing by livestock, but also changes in timing and frequency of fires and the disappearance of the indigenous large herbivores may be of significance. Temporary droughts in interaction with other disturbances may contribute, and climatic fluctuation or change may cause slow changes over long periods. What the anticipated global climatic change will mean for these areas, particularly concerning aridity, is not known.

The basic mechanism in bush encroachment is increased water availability for woody species as the grasses decrease their uptake as a result of damage from the overgrazing. This seems logical in a resource limited system with a balance between grasses and woody species. It can also be described as grasses being negatively and woody species positively correlated with herbivore density, at least if the herbivore is a mixed feeder or preferential grazer.

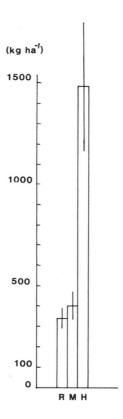

Fig.1: Total mass +- Standard Error of shrubs after a 5 year grazing experiment with no grazing (R), moderate grazing (M) and heavy grazing (H). (From Skarpe, 1990.)

The correlation is, however non-linear. There seems to be little change in the ratio between woody growth and grass under moderate grazing intensities, and pronounced changes in overgrazing situations (fig. 1). Bush encroachment may sometimes be a slow, gradual process, but often it is quick, and can build up more or less impenetrable thickets in less than 10 years. The reverse process is slower, and its mechanisms less known. There are examples where the thicket vegetation, once established, has remained for 30-50 years, even if the animals were removed. After that period the shrubs died fairly sudden without resprouting (van Vegten, 1981; own observations), and a grassy vegetation developed (Walter, 1954). There are a number of theories on grazing eco-systems with alternative equilibria, discontinuously stable systems and systems with stable limit cycles (Noy-Meir, 1982) which might offer an explanation of the observed behavior (fig. 2).

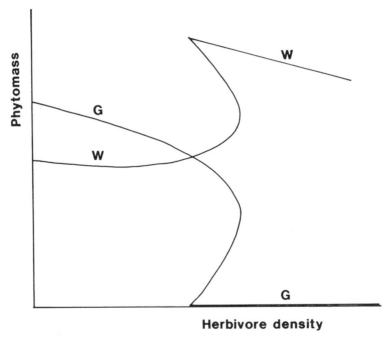

Fig. 2. *Response of steady-state herbaceous (G) and woody (W) biomass to herbivore density according to the model by Noy-Meir (1982).*

The above observations are all concerned with mixed feeders or preferential grazers. The tendency to bush invasion seems to be less with preferentially browsing animals, like camels, goats and indigenous browsers.

Physiognomic changes in savannas can also include the virtual disappearance of woody species, often caused by excessive cutting, cultivation - particularly with deep soil treatment - or frequent and fierce burning. In addition the regrowth is often destroyed by intensive browsing and by extreme micro-climatic conditions on the bare soil surface. The disappearance of trees may also be a response to decreased soil moisture, e.g. a lowered ground water table. Further, overgrazing in itself leads to poorer soil moisture conditions by compaction and/or capping of the soil and thus decreased infiltration rate and increased run off. Also short cuts for water infiltration along stalks and stems and root channels are lost as vegetation, particularly the grass sward, is damaged. Such effects actually makes the vegetation climate drier under overgrazing, without changing the meteorological macro-climate.

Changes in the Herbaceous Vegetation

There are few records on long-term changes in the composition of herbaceous vegetation of savannas or grasslands. Early travellers, who were not botanists, rarely give records on species level, and no form of remote sensing can accurately separate species of herbs. Anyhow, in southern Africa, large changes in the herbaceous vegetation have taken place well within living memory, and can be traced by direct information by local hunters, herders or ranchers - often even with names of species or species groups. There is also some information available from recent range research.

A problem in short-term vegetation studies is to distinguish between temporary fluctuations and long-term change. The savanna eco-systems are highly resilient, and they undergo profound changes in response to short-term disturbance, e.g. a temporary drought or heavy grazing, but later return towards a kind of equilibrium.

These fluctuations consist of continuous changes in absolute and relative abundance of species basically within a given set of species. The plants suffering most from the particular disturbance decrease most, and those suffering least decrease less or may even increase. The vital attributes determining the reaction of a particular species (or population or individual) may be of morphological, chemical, physiological, ecological etc. character. In a grazing system, for example, palatable species with a large proportion of their green mass within reach of the animals may suffer from heavy defoliation, whereas unpalatable, poisonous, prostrate or extremely thorny species may do better. Plants are rarely killed by the grazing itself, but get their competitive ability reduced, and are subsequently outcompeted by less grazed and/or less sensitive species.

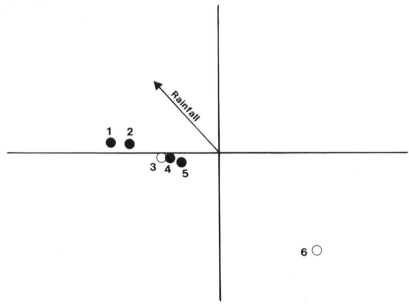

Fig 3: Ordinate diagram with vegetation types (1-6) situated according to their species composition. Rainfall increases in the direction of the arrow. Vegetation types 1, 2 and 3 are common in the same geographical area, 1 and 2 are not severely disturbed, whereas 3 is the result of severe overgrazing. Types 4, 5 and 6 occur in the same geographical region, with somewhat drier climate than the previous one. While 4 and 5 are reasonable undamaged, type 6 is overgrazing induced. (Canonical Correspondence Analysis; from Skarpe 1986).

Under severe disturbance or the co-occurrence of two or more disturbances, there is a definite change in species composition in that the least adapted species go extinct, and better adapted species, in many cases uncommon before the disturbance, increase. Under heavy grazing, for example, there is a change towards more and more ungrazeable species until a vegetation has developed, that is less sensitive to grazing and produces little fodder. The net production of the vegetation does not necessarily decrease, but the plant community changes, and will eventually have few or no species in common with the original one.

As mentioned above, overgrazing often results in drier growth conditions. This can be clearly seen from large scale vegetation studies, where overgrazing-induced communities may contain species otherwise found under drier climate (fig. 3). Another indirect effect of grazing is a speeding up of the nutrient circulation, and, relatively, improved nutrient access in the surface soil. This may also influence species composition, particularly in the vicinity of villages, watering points, kraals etc.

A few field observations indicate that vegetation changes following overgrazing may be stepwise rather than continuous, or be triggered by a temporary drought. Vegetation studies finding fairly well separated plant communities may also indicate the existence of more or less discrete stages (Skarpe, 1986)). Sometimes these can, with greater or lesser accuracy, be related to different kind or degree of disturbance.

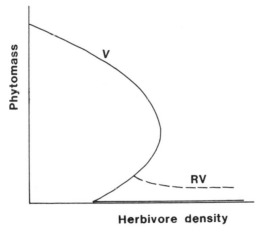

Herbivore density

Fig. 4. Response of steady-state plant biomass (V) to herbivore density; with ungrazeable residual vegetation (RV). (From Noy-Meir, 1982).

If overgrazing ceases, the vegetation may revert to its previous condition, or it may remain in the degraded stage depending on internal equilibrium conditions or changes in soil properties. Often the standing crop and, perhaps, the production quickly increases, but the change in species composition is slow or lacking. A large proportion of dry rangeland, e.g. the Sahelian savannas with annual grasses, may consist of such degraded vegetation (Cissé, 1986). As long as the disturbance is not such that no available species possesses the attributes necessary to survive, it is not likely that the vegetation will disappear altogether. This agrees with theories on equilibrium conditions in grazing eco-systems, e.g. the discontinuously stable model by Noy-Meir (1982) (fig. 4), presuming a non-grazeable residue vegetation - even if nothing is stated about its nature.It is now fairly clear that the threat of "desertification", understood as irreversible collapse of the vegetation cover, has been over-emphasized in the past (cf. Forse, 1989). The main problem is the degradation of vegetation around population centers rather than advancing dunes along the desert edges. The problem is not less serious, but it is different from what has been presumed, and it requires different measures. The emphasis on spectacular methods to more or less physically hinder the wandering deserts must to a large extend be changed to less popular efforts for stopping or reducing the overutilization of vegetation and soils - in order to maintain or raise the production of utilities for humans.

References

Cissé, A.M.: 1986. Dynamique de la strate herbacée des pâturages de la zone sud-sahélienne. Landbouwuniversiteit Wageningen, Wageningen.

Forse, B.: 1989. The myth of the marching desert. New Scientist 4: 31-32.

Noy-Meir, I.: 1982. Stability of plant-herbivore models and possible applications to savanna. In: B.J. Huntley & B.H. Walker (eds.): Ecology of tropical savannas, pp. 591-609. Berlin.

Skarpe, C.: 1986. Plant community structure in relation to grazing and environmental changes along a north-south transect in the western Kalahari. Vegetatio 68: 3-18.

van Vegten, J.A.: 1981. Man-made vegetation changes: an example from Botswana's savanna. National Institute of Development and Cultural Research, Gaborone, Botswana.

Walker, B.H. & Noy-Meir, I.: 1982. Aspects of the stability and resilience of savanna eco-systems, in: B.J. Huntley & B.H. Walker (eds.): Ecology of tropical savannas, pp. 556-590. Berlin.

Walter, W. & Volk, O.H.: 1954. Grundlagen der Weidewirtschaft in Südwestafrika. Stuttgart.

An Environental Monitoring Center in Senegal

Peter Frederiksen

Introduction

The Centre de Suivi Ecologique is a United Nations Sudano-Sahelian Office (UNSO) project financed by DANIDA and administered by the United Nations Development Programme, Office for Project Services (UNDP/OPS).

The project's major purpose is to create an institution responsible for the development and implementation of environmental monitoring capabilities in Senegal. The background was that a number of researchers working in the Sahel had suggested that the repeated Sahelian droughts had initiated a process of irreversible environmental degradation caused by overgrazing leading to a destruction of the grazing potential of the pastures and a development towards desertlike conditions severely affecting the life of the herders in the area. The conclusions were repeated in the media and with time, the Senegalese government and the donor community were largely convinced that a major effort had to be made.

The environmental degradation was, however, never quantified nor really proven. UNSO therefore started the present project to find out whether the supposed environmental degradation took place or not in order to define the appropriate development strategy for the Sahel.

The General Approach to Monitoring

The project has taken the approach of working with a multidisciplinary team in a computerized environment. The environmental monitoring is broadly defined as a spatial analysis of human production systems.

The project staff is composed of 35 persons, of which 2 are national and 5 international UN experts. Organizationally, the personnel is composed of a chief technical advisor and a national counterpart, technical advisors in ecology and geography, a series of project leaders within different disciplines directly accountable to management and the support staff (computer scientists, librarians, secretaries, drivers, etc.), whereas the resource experts maintain contact with the users.

From a technological viewpoint, the center has emphasized remote sensing, systematic reconnaissance flights and fieldwork in the data collection phase, because this particular combination permits monitoring of the environment at regular intervals and on different scales, including the national. The data analysis and interpretation phases are done in an entirely computerized environment including satellite image processing systems, point information management systems for the systematic reconnaissance flights and in general software capable of handling maps, tables and satellite images in a geographic information system.

The remote sensing activities utilize different types of imagery from high to low spatial resolution according to the environmental parameter. For rainfall measurements satellite imagery is required each hour, and the only satellite available for this purpose is Meteosat, which gives information in the required thermal band on a 25 square kilometer basis. For the vegetation and bushfire studies, where several images are required per decade and where a given combination of data is necessary to measure the growth intensity of the vegetation, NOAA-AVHRR satellite imagery with a 1 square kilometer resolution is needed. Both Meteosat and NOAA-AVHRR imagery permit low cost, high-frequency monitoring of Senegal. Landsat MMS, Landsat TM and SPOT imagery only cover a minor part of the country, less often and they are much more costly. They are particularly useful for detailed studies at e.g. lower administrative levels or for calibration of NOAA-AVHRR imagery because of their high spatial resolution - 0.6, 0.09 and 0.04 hectare, respectively. The University of Copenhagen's Department of Geography has provided essential support by supplying the CHIPS image processing software. Without this, the center could not have established itself so rapidly in the Senegalese environment.

The systematic reconnaissance flights were originally used for animal counts in northern Senegal, but they have later been found useful in central Senegal in relation to measuring agricultural land. The principle is that a plane flies at low altitude and speed along oriented flightlines separated by a given distance. Two observers observe and count animals in strips or bands on either side of the plane, and they also operate a camera taking vertical photography of the landscape. Each observation is given geographical coordinates, which makes it possible to acquire information on the number of animals and the relative cover of bare soil, natural grass and tree vegetation, agricultural fields and burnt areas. The data are then processed in the geographical information system, where they are reorganized according to the needs of the user, e.g. administrative, ecological or socio-economical units, and then combined with data from other sources, such as the satellite image data, field counts of animals at mechanized wells or other water resources.

The geographical information system is a software package which makes digital processing of geographical information, such as satellite imagery and maps, possible. A map can be transformed into digital format by digitizing it; it can be displayed and each map unit can be associated with an environmental parameter. For instance, the soil map of Senegal is digitized, and each soil polygon or soil map unit is registered in a table which gives its texture, depth, etc. It is then possible, if the appropriate algorithm exists, to associate these data with e.g. rainfall data and to model the development of the plant available soil water. Another example is the green biomass production data from satellite imagery, point information on cattle from the systematic reconnaissance flights, field information on socio-economic regions, that can be combined to produce a map of grazing intensity - a useful tool in rangeland management. Another advantage is that maps can easily be corrected and updated, which is particularly important in a monitoring center, where new geographical information is continuously added.

The Environmental Parameters Monitored

The environmental monitoring has focused on the mapping of vegetation (green biomass production, vegetation phenology, cultivated land and bushfires), water (rainfall, lake areas, flooding and mechanized boreholes) and animals (density), and socio-economic studies of the pastoral production system. The results have been presented in technical bulletins, as visual demos, in meetings with government agencies, at trade fairs and in the public media.

The first major issue the center addressed was the ecology and management of the pastoral ecosystem in northern Senegal among others related to the supposed process of irreversible environmental degradation. The green biomass production of the pastures had been significantly reduced in the years of drought, and it was generally accepted that they would probably not return to the pre-drought production levels, even with a return of "normal" rainfall. It was furthermore thought that the grassless landscape around the mechanized boreholes in the dry season was a result of overgrazing, and that as a result the production around the borehole was lower than in the surrounding landscape. It was therefore generally accepted, that the establishment of boreholes combined with the drought had induced overgrazing and irreversible environmental degradation.

The green biomass production maps produced by the center in 1987, 1988 and 1989 showed, that with a much higher (or "normal") rainfall in 1988 and 1989 compared to 1987 the green biomass production returned to pre-drought levels, and that there was no statistically significant increase or decrease in green biomass production with increasing distance from the borehole. In some cases the biomass could be higher, in some cases lower, at the borehole than, for example, 10 km away. The hypothesis of irreversible environmental degradation in terms of green biomass production could therefore not be proven.

The tree cover and the palatability of the pastures did change, however, due to fuelwood cutting and intense grazing, and in that sense the establishment of boreholes decreased the quality of the pastures around the boreholes. These zones, however, only cover less than 1 % of northern Senegal and are thus already insignificant in relation to the total extent of the pastures. The hypothesis of environmental degradation can therefore not be confirmed for northern Senegal.

In order to know whether the available biomass was sufficient to feed the cattle the center carried out animal counts during the systematic reconnaissance flights. They were carried out some time into the dry season when most of the cattle were organized around the boreholes with less movement from one borehole area to another thus avoiding double counting of cattle as much as possible. The results showed that in years of below average rainfall (1987) grazing pressure was only high in northernmost Senegal, and in years of normal rainfall (1988) there was twenty times as much available pasture as required by the cattle and other domesticated animals. The hypothesis of overgrazing in northern Senegal can therefore not be confirmed.

The government policy is to integrate the herders in the national economy in order to reduce meat imports. The center consequently studied the regional socio-economy. The weak or non-existent institutional links between the herders and the government, and the resistance of the herders, makes any government policy difficult to implement. Furthermore, production is hindered by the poor organization of herders.

The conclusion to be drawn then, is that since pastures and water are the most important natural resources for the animals and both are in ample supply in years of normal rainfall, there are no major environmental constraints on animal production within the pastoral ecosystem except in years of extreme drought. The main problem is of socio-economic origin, a problem that may prove very difficult to solve.

The center was therefore faced with the fact that an economically unimportant region did not have major environmental problems and that the socio-economic situation in this region might be difficult to change. And why change the functioning of the system if the herders are not interested?

Concurrently, it became evident that the techniques applied at the center could be used in the economically more important rainfed and irrigated agricultural production systems in central and southern Senegal. The center thus extended its operations on a nationwide scale, thereby also achieving some economies of scale. The center had built up an infrastructure of computers and other hardware and invested heavily in consultants and staff training in order to meet the demands of large-scale tasks. It seemed difficult to justify the presence of such a structure in a developing country, if only a minor part of the potential in the project was used. A reorientation towards economically more important sectors of the national economy therefore took place, a larger user group of center products was created, and the center became a more "profitable" project in terms of impact on government agencies. The center now also works with bushfires, water surfaces, etc. and expects to contribute significantly to the agricultural statistics and to land use planning and management in the future.

Bushfires are in official government policy considered an environmental hazard that should be limited as much as possible, but their causes and particularly their effects and spatio-temporal distribution are not well known. Government agents ignite early fires when the pastures are still somewhat humid, whereby the fire temperature, areal extent and effect on vegetation is lower or more limited as compared to later fires occurring when the vegetation is completely dry. They also combat later fires (this is done from a conservationist point of view). The rural population follows the same practice - as soon as the rainy season is over, extensive bushfires unrelated to government activities appear. They are often ignited to improve the grazing grounds, kill ticks and snakes, to improve the trafficability of the landscape, to fight locust attacks, to increase the resin production of trees for gum collectors, etc. The long term ecological effects of the bushfires is largely unknown.

The center has attempted to map burnt areas by using the NOAA-AVHRR satellite imagery on a monthly basis. Although the method needs improvement, it has, for the first time, been possible to obtain a

41

good approximation of the extent of burnt areas. The data indicated that up to 40,000 square kilometers had been burned from October 1988 to January 1989 - ten times as the official figure.

The figures compromised the government agency responsible for the official figures, and severe tensions arose between the center and the agency. At present the situation has improved, and an agreement has been signed according to which the center will supply areal statistics on bushfires, based on remote sensing techniques in order to improve the official bushfire statistics.

Future bushfire investigations carried out by the center may be placed within the socio-economic sector in order to produce a better understanding of why the local population ignites fires. It is quite possible that the motivation varies between different human production systems such as agriculture, nomadism, gum collection, hunting and charcoal producers, as well as from one physical geographical region to another. An already on-going activity is the cooperation between The University of Copenhagen and the center in developing a more precise bushfire mapping methodology.

Rainfall is a highly variable parameter in Senegal as in the rest of the Sahel. Even though Senegal has one of the densest network of rain gauges in the region, the spatial variability of the rainfall is so high that the rain gauge data cannot be used to produce a reliable rainfall map of the country. The University of Reading had for some years used Meteosat data to estimate rainfall for the Sahelian region, and their method was adapted to Senegal for the 1989 rainy season. Every ten days, each month and for the rainy season as a whole, rainfall maps were produced on a 25 square kilometer bases, a much higher resolution than obtained with the rain gauges alone. These data were requisitioned by the Meteorological Office under an agreement with the center.

The possibilities of mapping lake areas, flooded areas and irrigated agriculture in the Senegal river valley by using NOAA-AVHRR have been explored. The preliminary results indicate that NOAA-AVHRR imagery may possibly be used for lake area and flood monitoring if calibrated with higher resolution satellite imagery, and that further developments of the methods are required before the method eventually can become operational. The prospect for mapping irrigated agriculture by using NOAA-AVHRR imagery is dim as high resolution imagery such as Landsat TM or SPOT seems to be required.

Due to the importance of NOAA-AVHRR imagery in the vegetation, bushfire and water studies, the center is building an archive of geometrically resampled images of the UTM projection to make them comparable with the topographical maps of the country. This supplies the center with a historical dataset for the 1980'es and at the same time makes the incorporation of the imagery in the geographical information system of the center possible.

The center has not yet attempted to measure agricultural production but it will probably be a major field of investigation in the future, developed under a common agreement with the Ministry of Agriculture. However an appropriate methodology will have to be developed first, the problem in the senegalese agricultural landscape being the small fields with trees on the fields separated by small villages and non-cultivated areas. This makes it difficult to use NOAA-AVHRR imagery, because so many land cover types will be present within each pixel. A second problem is, that green biomass production is not necessarily well correlated with grain production. It seems that elements which could be used include high resolution satellite imagery to stratify the agricultural landscape into regions, systematic reconnaissance flights with cameras and videorecorders to obtain statistically sound data on land use within each region, combining this with field studies of grain production in sample areas. Several universities and international research institutions would have to participate in the development of a methodology, but once operational it could significantly help in improving the presently imprecise agricultural statistics and thus produce a more precise basis for planning the national economy and rural development.

The Future Role of the Center

Being an institution building project and thus a new phenomenon in the Senegalese government structure, the center has had to define its own role, to become known, and it has had to have a positive impact on decision-making. If not, the center could not justify its existence. A considerable and quite aggressive marketing effort was launched, so intense that several agencies felt threatened by the center because they regarded it as highly competitive and potentially dangerous in terms of its eventual capability to attract attention from decision-makers and donors, and thereby money and power. On the other hand, decision-makers and government technicians became aware of the utility of the center. At the same time the center became more assertive of its own capabilities and a working environment of mutual confidence was generated between the center and several, key government agencies, who saw a possibility of expanding their knowledge within their particular fields if supported technically in specific areas by the center.

At present the center's major weakness is its institutional attachment to the Ministry of Nature Protection, in the Board of Forests, Water areas and Hunting, which is mainly concerned with conservation and not with environmental monitoring, planning and management - the center's major fields of interest. It has been suggested that the center should become part of the Ministry of Planning, or that it obtain a parastatal status between a government agency and a non-governmental organization.

The future management on the Senegalese side might be required to have a background in business management instead of ecology in order to orient the center towards a more realistic budget and more realistic user-driven activities. Each activity will probably have its own contract personnel and budget so as to relate the project leader in a very direct manner with project management. It is furthermore expected that the UN international/national expert and the civil servant/UN national expert ratios will decrease in order to emphasize the national participation and to increase both the technical and theoretical know-how of the center.

Animal Counting in the Northern Senegal (Ferlo). A Comment on the Reliability of Quantitative Livestock Data.

Kristine Juul[1]

One of the major obstacles encountered by rural development planners in Sub-Saharan Africa is the lack of reliable statistics. Serious efforts are being made to collect data and to organize databanks but many of the results remain inaccurate and so must be used with extreme caution. This is particularly true of livestock data collected in the sahelian zone because of the absence of appropriate methodologies which take the fact that herds are usually large, of mixed species and frequently cover large distances, into account.

Official statistics are often of far poorer quality than their users may suspect. Too often data is "fudged, cooked and manipulated by officials at higher levels, the main purpose being to ensure that the trends will be found satisfactory and convincing to those with still greater authority, as well as to compensate for presumed biases, which are usually unrelated to the realities of tropical conditions"[2]. Nevertheless, studies of recent developments in pastoralist societies in the Sahel often present these dubious quantitative data in order to support their different hypotheses. Reports emanating from the different donor agencies often contain extensive data proporting to show such information as the evolution of livestock numbers over a given period of time, the percentage of livestock that perished during various periods of drought, or calculated stocking rates, etc. However useful these quantitative measures might be, the following experiences from the execution of a

---

1. An extended version of this article exists in French and was produced in collaboration with Daniel N'Decky, Direction de l'Elevage and Oussouby Touré, Centre de Suivi Ecologique. The author of the present article is entirely responsible for its contents.
2. Polly Hill, "Development Economics on Trial", Cambridge 1988, p.33.

45

cattle counting survey in Northern Senegal give an impression of the conditions under which such basic quantitative data is collected and thus of its reliability.

In Senegal, the most commonly used data on livestock numbers and stocking rates are provided by the veterinary authorities of the Direction de l'Elevage. These data are estimates derived from the number of cattle vaccinated during compulsory (and free) campaigns carried out annually over the entire country. Unfortunately a number of significant sources of error in the techniques of data collection and manipulation exist:

Firstly, a significant portion of the cattle are not presented for vaccination by the herders. This is particularly true of calves less than 6 months old, and animals 8 years or older, since they are considered to be immunized, and milch cows in order that they are not subjected to stress. Estimates of herd size which include those animals not presented are made by multiplying the number presented by a constant.

Secondly, there is no compulsory vaccination for sheep, goats, horses and donkeys, so that estimations for these species can only be very approximate estimates made by the local veterinary agents.

Finally, corrections to the vaccination data are made at several different levels by both local and regional authorities in such a manner that data becomes increasingly distorted, the further away from reality it gets.

In order to help improve the reliability of official statistics concerning livestock numbers and distribution, the Centre de Suivi Ecologique (The Ecological Monitoring Center) in Dakar (CSE) has carried out several systematic reconnaissance flights since 1987. The method involves flying parallel flight lines spaced at 5 km intervals and of counting all livestock that fall within defined 150 m wide bands either side of flight lines[3]. Data are allocated to individual 1 km² units by dividing flight lines into observation units according to a timed period of flight, the data itself being expressed as livestock (head)/km². Such a presentation of data allows the variability of densities to be tested for given mapping units.

As a further tool in the assessment of livestock estimates, a simultaneous ground campaign was carried out in the region of 14 deep

---

3. Faye, Marks & Prévost, 1989: "Enquêtes sur les effectifs de bétail et leur distribution dans la moitier nord du Sénégal 1987-89".

wells in the agropastoral zone of northern Senegal (the Ferlo) during
April 89 (figure 1).

Deep wells included in the animal counts

1	Amali	8	Mbiddi
2	Bouteyni	9	Namarel
3	Belel Boguel	10	Niassanté
4	Ganina Erogne	11	Tatki
5	Kothiédia Haïre	12	Tessékré
6	Labgar	13	Widou Thiengoly
7	Mbar Toubab	14	Yare Lao

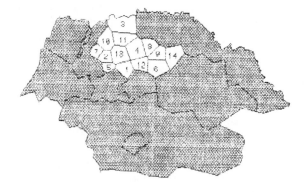

Having established these basic facts enabled us to collect considerable
supplementary information regarding herding systems in the area (e.g.
watering patterns, exceptional and seasonal livestock movements,
reactions to bushfires, cattle diseases, as well as the main problems
related to the functioning and management of the deep wells). Furthermore it gave a vivid impression of the conditions under which such
data must be collected.

The enumeration of livestock was carried out by two teams working
simultaneously at 7 wells. On arrival in the zone served by a deep
well, contact was made with the local well committee in order to
inform the herders of the purpose behind the counting exercise and to
gather basic information on watering patterns and the functioning of
the deep well itself. According to the size and importance of the well,
4-6 enumerators were placed at strategic positions on the main cattle-
paths leading to the cement troughs constructed around the wells. The
counting took place from 7 a.m. to about 7 p.m. over a 2 day period -
in order to include the numerous animals that were watered only

47

every second day. In cases where part of the livestock was watered at traditional wells, extra days were added to the programme to ensure that these animals were included in the survey.

Unfortunately, the extreme variability of the sahelian pastural system is considered to have significantly influenced the reliability of the data collected at the wells. The main problems encountered were:

1) There was a somewhat ambiguous relationship between the enumerators and the local population since herders in general are extremely reluctant of having their animals counted on an individual basis. This reticence is partly due to fear of a return to former cattle taxing systems[4] and partly to the traditional beliefs that herd counting can lead to increased mortality. However, a few cases of threatened violence apart, this turned out to be a less important problem than had been envisaged, since the counts were made on a collective basis with neither owner nor cattle brands being registered. Furthermore the count was carried out in collaboration with both the well committee and the local veterinary agent.

2) The different habits in watering and grazing livestock adopted by different households inevitably led to some herds being counted twice. As the pasture near the deep well get scarcer as the dry season progresses, most herds switch from daily watering to watering every second day. This enables them to exploit more distant pasturelands during the day the cattle are not watered. Nevertheless, cattle owned by families living in the proximity of the deep well continue drinking once or even twice a day. For obvious reasons, it was impossible for the team to distinguish the different herds. Therefore a pragmatic strategy was adopted which sought to remove from the count second and subsequent visits to the well of the same animals. This was achieved by subtracting from the total two-day counts an estimate, made by the well committee, of the number of cattle considered to have visited the troughs each day[5].

3) The problem of double counting was even more pronounced when it came to donkeys, horses and dromedaries. Donkeys and horses normally graze freely in the general vicinity of wells, visiting the

---

4. Taxation of cattle was abandoned after the drought of 1972/73.
5. These estimates were very rough, ranging from none over one third to half of the animals present.

troughs several times during the day. In the early mornings up to 700 donkeys were observed in the immediate vicinity of the well. A proportion of these would later return hitched to the carriages transporting water to the compounds. Dromedaries also complicate matters since they need watering only once or twice a fortnight, reducing significantly the credibility of our method of estimating them. The problem was made worse by the decision of the Senegalese authorities, taken in November 1988, to expel from Senegalese territory the numerous herds of dromedaries owned by Mauritanians[6]. Some herders ignored the decree and so avoided taking their animals to the well during the counting sessions. Well committees were equally reluctant to discuss this subject, as influential Mauritanian herders seemed to enjoy the protection of certain board members in order to remain in the zone.

4) The mobility inherent in pastural systems creates an extreme fluctuation in the number of livestock present in an area at any given moment. Even though most herders in the Ferlo now practice a more or less semi-sedentarized production system, their mobility being reduced to a radius of 15-20 km from a particular well, some agropastoralists still practice seasonal movements on a large scale. Several example can be presented here:

Seasonal movements may be practiced by herders towards more urban regions to facilitate the sale of milk, butter, sheep and goats. Others practice seasonal migration to cultivate along the banks of the Senegal River and bring with them some of the milking cows. Health considerations may also bring about temporary migration, either to areas where the pastures, rich in salts,[7] provide a cure for certain deficiency diseases, or to flee from areas infected by livestock diseases. Differences in the watering taxes equally push some herders to pass the dry season in zones with lower tax rates. Furthermore a number of exceptional cattle movements occur, for example due to the large bushfires of

---

6. According to this decree, herders were not allowed to have more than 1 female and 2 male dromedaries.Only a small minority of Senegalese herders have dromedaries, the target of the operation being the numerous Mauritanian herds present in northern Senegal since the drought of 83/84. The expulsion of the dromedaries was one of the factors leading to the violent riots between Senegalese and Mauritanians in April/May 1989. After the riots, the herds of dromedaries have practically disappeared from the zone.
7. Along the banks of the Lac de Guiers, the ancient Ferlo Valley, or the Senegal river.

november 1988, which spread over more than 2,500 km². At certain deep wells in the center of the fire zone[8] approximately half the herds had left, presumably to return by the first rains[9]. No reliable information exists on the volume of any of these movements.

5) A proportion of the herds grazing in the vicinity of the deep wells never enter the "counting-zone" as they are watered either at small waterholes dug in depressions of the ancient Ferlo Valley or at compounds a few kilometers away from the deep well. The latter is an effect of what might be called the "small technological revolution": the use of tractor inner tubes transporting up to 800 liters of water on donkey carts. By watering parts of their herd in their compounds, herders manage to make considerable savings on the taxes paid for watering, as well as raising herd productivity by making it possible for the animals to reach more distant pasturelands, and sparing them from long trips to the wells. This method of watering is essentially practiced for the youngest and oldest animals and particularly goats and sheep. However, certain households that possess several donkey carts manage to water a considerable part of their herd at the compound. At certain deep wells,[10] where this practice was especially widespread between 100 and 200 donkey carts were daily observed transporting water. This of course represents a serious obstacle to the whole concept of counting at deep well level.

Comparison of Different Sources of Information

The following table presents some examples of the considerable variations encountered when confronting different types of information at deep-well level[11].

---

8. Tatki, Ganina and Mbiddi.
9. It was interesting to note the number of herds (particularly of sheep and goats) which had stayed or already returned to the zone several months before the rainy season.These households and herds had installed themselves around the few fringes of grass untouched by the fire, arguing that the ashes left by the fire had a curative effect on certain deficiency diseases such as botulism.
10.Labgar, Namarel Widou Thiengoly, Yare Lao and Belel Boguel. As the well at Tatki was out of order during the counting campaign this might explain the large number of carts visiting the well of Belel Boguel.
11.Estimations from all 14 deep-wells can be found in the original version of this paper.

Table 2. Estimations of livestock at different deep-wells

	Cattle	Small Ruminants	Donkeys	Horses	Drome- daries
Bouteyni					
Vaccination campaign 88/89	2,814	-	-	-	-
Animals paying water tax	1,189	4,069	181	16	-
CSE campaign (2)	4,688	7,578	1,164	722	-
Ganina (3)					
Vaccination campaign 88/89	1,109	3,900	-	-	-
ISRA 87	2,000	15,000	861	83	49
ISRA 88	1,500	17,700	903	89	63
CSE campaign	3,341	4,403	630	63	1
Labgar					
Vaccination campaign 88/89	6,200	-	-	-	-
Veterinary's estimate	7,750	10,226	-	-	-
SODESP's estimate (4)	10,000	-	-	-	-
Animals paying water tax	4,911	8,205	-	-	-
CSE campaign	9,899	9,882	-	-	-
Niassanté					
Vaccination campaign 88/89	9,363	-	-	-	-
Animals paying water tax	2,400	989	36	15	-
CSE campaign	9,945	18,381	2,402	545	-
Yare Lao					
Vaccination campaign 88/89	5,629	-	-	-	-
Animals paying water tax	1,215	5,586	-	-	-
Well committee estimations	10,000	7,000	-	-	-
SODESP estimations	7,000	10,000	-	-	-
CSE campaign	4,843	7,110	-	-	-

(2) Institut Sénégalais de Recherche Agricole.
(3) The well-committee in Ganina had lost the book where the tax accounts were listed. Information was thus un-available in this instance.
(4) Société de Developpement de l'Elevage dans la zone Sylvo-Pastorale. A state-run cattle development project based in Labgar.

The extreme variations shown in the table makes it impossible to extract any reliable trends. At almost all water sources considerable differences can be observed, the most impressive discrepancies occurring at Labgar and Yare Lao. Part of these variations can be explained by variations in the seasons and years when the estimations where made. Nevertheless even estimations derived from the same period differ significantly and the distortions observed between different sources show an extreme variance from one deep well to another.

First of all it should be mentioned that the numbers derived from the veterinary services are only raw data and have not yet been corrected for the animals not presented for vaccination. Unfortunately the data coming from the well committee are even more dubious, considering the reluctance of herders to be assessed for water taxes and the large number of animals being watered at compounds. In fact most well committees determine the taxes to be paid by herds on a very rough estimate, often based on individual agreements between herders and the committee. Actually, most well committees face severe budgetary problems when it comes to paying for oil, fuel and the salary of the mechanics since only approximately half of the animals are actually accounted for. Such chronic budgetary deficits contribute to frequent mechanical problems and prolonged water cuts.[12]

When considering the figures originating from ISRA, our experiences in the field leads us to some methodological criticisms. According to the ISRA method the animals watered at compound level (thought only to be calves under 6 months of age) can be estimated on the basis of the number of female cattle presented at the troughs and the seasonal distribution of confinements. This is unsatisfactory since a proportion of the females are watered at the compound as well and, more importantly, the practice is much more widespread for the small ruminants.

Comparisons with the Aerial Survey 1989

Estimates obtained by the aerial sampling for small areas, such as individual wells, are not sufficiently accurate to be presented here. However their degree of significance increases as the area surveyed and

---

12. This negligence towards one of the principal factors of pastoralist production is a reaction to the withdrawal of the state from the running and maintenance of the wells. Until 1984 when the management of the wells was handed over to well-committees elected from amongst the herders, the state covered all the major expenses of the wells.

the number of observations made increases.[13] Thus only data from the area served by all 14 deep wells is presented (table 3).

Table 3. Comparison between the data of the CSE ground count and its aerial survey

	Cattle	Small Ruminants	Donkeys	Horses	Drome- daries
CSE ground counts	84,895	129,220	18,323	4,345	747
Aerial survey 1989	66,282	137,835	3,766	3,013	753

The number of cattle estimated during the aerial survey gives a total approximately 20% less than that of the ground estimates. This is certainly due to an inherent underestimation of approximately 15% in the aerial survey data resulting from the fact that counts obtained from the aircraft when directly over wells are not included since the abnormally high concentrations of animals found there distort the total estimates to a greater extent than if they are excluded.[14] The values for the small ruminants are very similar. As the aerial survey is constrained neither by differing watering rhythms nor by watering in the compounds, the figures from the aerial survey were expected to be much greater. Two hypotheses are possible. Either it is due to the generalized under-estimations inherent in the aerial survey technique, or to an over-estimation of the number of livestock watered in the compound.

The estimates for donkeys vary enormously. Such a large variation certainly comes from methodological weaknesses of both assessment techniques. On the one hand is the under-estimation made by the aerial survey while on the other is the problem of double or triple counting made on the ground.

Finally the aerial surveys give a fairly "instantaneous" impression of the localization of herds as they don't take into account the changing livestock movements. In certain areas this may have a quite significant influence, disturbing comparisons from one season/year to the next.

---

13. Faye et al. 1989.
14. Faye et al 1989.

Conclusion

Many methods for estimating livestock numbers exist, all with their own inherent failings. The confrontation of two different livestock counting methodologies used by the CSE rendered the shortcomings of both types of data collection visible. The aerial survey is considered to give much better estimates over larger areas, and lends itself to statistical analysis of its significance. However it is inefficient over small areas and estimates are considered to be fairly inaccurate. Furthermore it takes no account of, for example cattle movements. The data collected at ground level provided important complementary information on livestock movements, watering patterns and the management of the wells, but still produced fairly inaccurate figures.

The necessity of further development work on a methodology to produce more reliable data seems obvious, but meanwhile available data must be submitted to extremely cautious use. Because of the unreliability of the data presently available, and of the difficulties in extrapolating from studies made on a grass-roots level to greater geographical regions, many statements on the evolution of livestock populations as well as their geographical localities tend to become axioms (i.e. the available statistics are too unreliable to prove them).

Finally, the idea that unreliable statistics are better than no statistics at all, might easily lead to dubious theoretical conclusions and mistaken strategies.

The Political Economy of Agricultural Development and Cereal Marketing in the Sahel - the Case of Burkina Faso

Mike Speirs

Synopsis

In my initial presentation of this research project (in August 1987), I put forward three hypotheses which have formed the framework for an analysis of the dynamics of cereal marketing and agricultural development in Burkina Faso:

1. The production, consumption and distribution of maize, millet and sorghum (the principal grain products of the Sahel) are influenced by a complex set of relationships between village-level cereal banks, private cereal traders and the parastatal marketing board (OFNACER[1]). Therefore, the roles of these agents involved in the marketing chain should be investigated with a view to determining the extent of conflict between the public and private sectors, and to assess which system smallholders and agricultural producers stand to gain from, in terms of improved income and food security.

2. It has been argued that state policies which aim to ensure stable food consumption levels and remunerative producer prices are the result of a series of "economic tactics" which combine local initiatives and national development objectives. Studies of the price structure of supply and demand, as well as non-economic factors affecting cereal marketing, show that policies which reverse discrimination against the agricultural sector must be implemented.

3. Finally, the impact of international and regional trading patterns and relative prices of cereal products (domestic and imported) affect the staple foods market in Burkina Faso, and the structure of comparative advantage which determines the competitiveness of "traditional" cereal

---

1. The "Office National des Céréales", established in 1971, is the grain marketing board in Burkina Faso.

crop cultivation in the Sahel. These factors are also influential in terms of the institutional structure of cereal marketing.

Through studies of the relevant literature, through "field work" research in collaboration with colleagues at CEDRES[2] in 1988, and through the production of three working papers in 1989[3], I have applied a number of theoretical models in the investigation of cereal policies in Burkina Faso. In this paper I will briefly summarize the approach which I have adopted, focussing on the elaboration of hypotheses concerning agricultural development strategies and cereal marketing, methodological issues, and the structure of the final report.

Efforts to devise and implement food and agricultural policy reforms in many African countries during the 1980s have attempted to come to terms with what has become known as the "food price dilemma", which arises from a conflict of interest between farmers who want higher prices for their products and consumers who desire access to low price foodstuffs. Associated with this contradiction is the argument that African governments have exhibited an "urban bias" in economic policy making, such that discrimination against the smallholder peasant-farming sector has become widespread as a result of price policies and marketing institutions which primarily benefit urban food consumers. Thus it has been repeatedly argued that the prices of agricultural products should be raised in order to encourage farmers to produce and market greater quantities of both food and cash crops, and that the marketing systems themselves should be liberalized through the withdrawal of state controls on imports, exports and domestic distribution, as well as the dismantling (or at least partial privatization) of the parastatal marketing boards.

The main theme which underlies this research project concerns the contradictions and compromises which are inherent in the relations between state institutions and agencies, and the structures of rural communities in the Sahel where markets play a significant role as focal

---

2. "Centre d'Etudes, de Documentation, de Recherche Economique et Social" de l'Université de Quagadougou.
3. These papers are:
   Peasants, merchants and the state - some reflections on cereals policies in Burkina Faso. (an initial survey of the issues) February 1989.
   The political economy of agrarian change and the revolution in Burkina Faso. May 1989.
   An investigation of food price policies, agricultural marketing and trade strategies, with reference to the cereals market in Burkina Faso and food security in the Sahel. August 1989.

points of commerce, exchange and social cohesion. Against this background, my investigation into the complexities of the processes of food and agricultural policy reform, using data gathered from reports and interviews in Burkina Faso, suggests a reformulation of the initial hypotheses as follows:

Firstly, while cereal marketing largely lies in the hands of private traders, the parastatal marketing board has an important role to play in maintaining adequate levels of food availability in areas of chronic production deficits, and amongst the poorest food consumers (whose "entitlements", or effective demand are limited). Cooperation between the private sector (including the "groupements villageois" which run cereal banks) and OFNACER is essential in order to ensure that peasant farmers are able to maximize their incomes through the sale of cereals, while simultaneously maintaining consumption levels during the "soudure"[4].

Secondly, while "getting prices right" may be an important part of an agricultural policy reform programme (in connection with balance of payments stabilization and structural adjustment), it is not the end of the story. Effective agricultural development strategies necessitate not only improvements in incentives to farmers, but also improvements in rural infrastructure, in a better provision of farm inputs and credit, and in the availability of consumer goods. In the context of predominately subsistence food crop production in Burkina Faso, these "non-price" factors are highly significant in term of agricultural productivity and output gains.

Thirdly, the emergence of a distinct trend towards increasing "food dependence" through cereal imports and food-aid raises several problems which are not easily resolved. Wheat and rice are steadily substituted for "traditional cereals" in the diet of the urban population in many West African countries. In order to make indigenous millet, sorghum and maize (and rice) cultivation competitive in comparison with the cheap imports, protection of smallholder producers through tariff barriers or other measures is one, disputed possibility.

A thorough assessment of these hypotheses concerning the cereal market in Burkina Faso and agricultural development strategies in the Sahel draws attention to several important policy issues, while at the same time highlighting a number of practical difficulties. For example,

---

4. The "soudure" is the "hungry gap" in the agricultural season during the rains and before the harvest when food supplies are at their lowest levels in the rural areas of the Sahel.

it is not easy to draw definitive conclusions about the relative effectiveness of the various "actors" operating in the cereal market, given the complementarity of their roles and functions. Similarly, estimating the precise impact of price policies in terms of total food crop output is rather difficult in an environment where fluctuating rainfall levels exert a decisive influence on production. Nor, to give another example, is the world market a neutral agent with respect to protective policies, since the possible effects of agricultural trade liberalization on the prices of wheat and rice may have a significant impact on living standards and food security in West Africa. Thus, one of the aims of this investigation is to indicate the range of factors which have a bearing on the form and content of agricultural policy reforms in Burkina Faso (and, where relevant, in other countries of the Sahel region).

Nevertheless, I intend to illustrate and the ramifications of these hypotheses through an examination of data gathered in Burkina Faso and contained in reports on food and agriculture in the Sahel. Why has the state incurred budget deficits through the operation of OFNACER, and how can this parastatal body be reorganized such that both cereal producers and consumers will benefit? Why do incentive price policies appear to have had little impact on the total marketed food crop surplus, and what are the determinants of production and marketing behavior in the subsistence sector? Is it realistic to envisage the creation of a protected regional cereal market in West Africa, given the high domestic resource costs of increasing the production of rice and coarse grains, as well as the effects of such production on the welfare of consumers?

The answers to these questions can only be found through lengthy and exhaustive data collection and analysis in the rural and urban areas of the Sahel. Although this work has only recently been initiated and few results are as yet available[5], my own research has been based on exploring various dimensions of these questions, to the extent that existing production, price, marketing and trade data can be used to identify constraints and bottlenecks in the structure and functioning of the cereals on different markets in different regions of the country at different times of year is a first step towards assessing the efficiency

---

5. Two long-term cereal market surveys are underway in Burkina Faso: coordinated by the "Projet Diagnostique Permanant" of CILSS (the permanent interstate committee for drought control in the Sahel), and as part of an "observatoire sur les systemes alimentaires" (at CEDRES, Université de Quagadougou).

and performance of the various marketing "agents". At the same time it is possible to use price and production data in estimating supply responses, although this exercise is rendered difficult by the lack of detailed information about the prices prevailing on "parallel" markets (as opposed to official prices fixed by the price commission and applied to all OFNACER transactions). Interviews with private traders, members of village associations responsible for cereal banks in zones of deficit and surplus production, and with representatives of OFNACER in Bobo Dioulasso, Quagadougou, and Quahigouya have been carried out in order to supplement the statistical data gathered from a number of published reports.[6]

Information about domestic production, marketing systems and consumption patterns in Burkina Faso can also be combined with descriptive analyses of the regional and international dimensions of the cereal market in the West African Sahel, in order to examine the implications of the trend towards increasing demand for rice and wheat (and rapid growth in imports since the 1960s). This also leads to an assessment of the various options which might be envisaged to enhance food security in the region, and to a discussion of the social and political implications of measures which might be introduced to "counteract the urban bias". An examination of the "populist" pro-peasant development strategy adopted in Burkina Faso since 1983 is particularly interesting in this context, since numerous conflicts and contradictions have arisen through attempts to "take the side of the peasants" and to promote investment in smallholder production. In a sense, the main problem which must be tackled through agricultural and rural development policies and programmes in the Sahel, is that with high rates of population growth and a rapidly deteriorating natural environment (characterized by deforestation, soil erosion, etc.), emigration from the rural areas has exacerbated low rates of productivity increase in agriculture, but at the same time there are few opportunities for labor absorption in non-agricultural activities. If food selfsufficiency and reduced cereal imports are considered to be important policy objectives, a series of measures will be required to increase the competitiveness of the agricultural sector, and to encourage income and employment generation in the rural areas.

In the final project report, a survey of various theoretical approaches to studies of food and agricultural policies in Africa will be combined

---

6. This information will be brought up to date during my visit to Burkina Faso in October 1989.

with an empirical analysis of agrarian change and cereal marketing in Burkina Faso. Thus, it is intended to apply the models which have been developed to explain the processes of agricultural transformation in the context of economic growth (including theories dealing with production, marketing and trade), to an assessment of policies which aim to increase agricultural output, productivity, food security and incomes in the smallholder farming sector. Some sections of this report have already been covered in the working papers (see note 3 above), but a revision of these documents will be necessary in order to distinguish between theoretical and empirical issues and to supplement previous analyses of cereal-market data (the "balance sheet" showing production, consumption and trade trends, and price series). Despite data deficiencies, the interdisciplinary approach adopted in this research project will enable me to present a set of conclusions which take, the economic, ecological, social and political obstacles lying in the path of cereal marketing reform and agricultural development in Burkina Faso, into account.

Research and Development in Eastern Sudan: The Red Sea Area Programme

Leif Manger

This paper is based on a talk presented to the 1990 Danish Sahel Work-Shop. In that talk, I gave an outline of the Red Sea Area Programme (RESAP), a joint research programme between the University of Bergen and University of Khartoum. My main aim was to present the major research problems within that programme and to show how research could be of applied value and tied to problems facing development planners in that particular region. In this paper I shall also spend some time discussing two other major points that seem important within the context of the current discussions between the Danish Sahel researchers. Firstly, as RESAP is also an example of research cooperation between two universities through which the aim is to combine academic competence building with applied research I shall also discuss our experiences in this field. Secondly, as NORAD and MDC (Norwegian Agency for International Development and the Norwegian Ministry for Development Cooperation) have been financing important parts of this cooperation some reflections will be made on how research institutions and development agencies can work together.

The Norwegian SSE-Programme

RESAP is part of the programme "Environment and Development in the "SSE-countries" (Sahel-Sudan-Ethiopia, called the SSE-programme) financed by MDC. This programme is a Norwegian effort to assist in the rehabilitation process after the African drought. It is a long term programme mainly focussing on Ethiopia, Mali and the Sudan, the first phase being the period 1986-1990.

The programme consists of three main components. 1) Support to international organizations (about 50% of the funding), 2) support to Norwegian non-governmental organizations and their programmes in the area (40% of funding) and 3) support to research for research collaboration between Norway and research institutions in the SSE-countries (10% of funding).

The overall aims of the programme are:
- to improve local food production and food security
- to improve the natural ecological base in order to develop sustainable production systems.

The objectives relating more specifically to the research part of the programme are:
- to develop research competence and research institutions' capacities relevant for improved development efforts in the relevant countries
- to improve action oriented research to support NGOs
- to improve Norwegian research competence on the region.

The Red Sea Area Programme: Research Problems

The Red Sea Region in Eastern Sudan presents us with several problems that are typical for the SSE-belt. A very arid environment with low and unreliable rainfall and human adaptations based on animal husbandry and cultivation. The herding of animals in search of pasture and water and cultivation in the rainy season (if there was enough rain), constituted a flexible adaptation that made it possible for the Beja, the dominant group in the area, to survive within a marginal environment. Certain developments however, tend to create imbalances in this system. Population increase is one such factor that is of importance. Others are e.g., the trend towards a more sedentary form of settlement among the Beja. The big city of Port Sudan, as well as smaller regional centers, are growing rapidly in population. This leads to more animals, especially goats, being kept around such centers, as well as a rapid deforestation due to wood collecting and charcoal burning. Secondly, new opportunities provided by urban employment in towns as well as schemes like Tokar and Gash, do not relieve pressure on the traditional adaptation and pastures by absorbing people in non-pastoral activities. They instead seem to increase pressure by allowing people to maintain a stake both in pastoral and non-pastoral sectors.

There is thus a situation in which a growing population of Beja have to deal with an increasingly marginal environment and the recent drought and subsequent disaster is a dramatic example of this. Given the conditions of a) an arid natural environment, b) a complex Beja social organization and c) a modern context of markets, urban based employment and mechanized schemes, it is reasonable to assume that planning and implementation will take place under extreme uncertainties, and that the risk of failure and undesired consequences are correspondingly high. Reduction of this risk factor is clearly dependent on an increase in the information planners have about how the population is likely to react to alternative development inputs and about what the likely environmental consequences of these reactions will be. Part of the research task within RESAP is therefore, to collect data on the basis of which we can analyze the main ecological and socio-economic mechanisms of the region in a way which allows for some understanding of the probable implications of alternative plans.

The basic problem of RESAP is to look at ecological and social change in time and space. Being an inter-disciplinary project RESAP approaches this problem from different perspectives. The basic disciplines represented within the programme are those of *botany*, *geography* and *social anthropology*. Obviously they look differently at processes of change and they define their problem fields in different ways. In the botanical work as well as that of physical geographers studies of changes in the natural environment are made. Through ecological studies changes over the last decades can be mapped using botanical field-work in combination with information obtained from air photos and satellite imagery. Different vegetation zones can be mapped and change producing factors identified. Through pollen studies more long term changes in nature can be uncovered. Physical geographical studies can help decide to what extent rainfall, topography, geology etc. affect the above trends, particularly as vegetation is dependent on the many rivers that start in the mountains and end in the Red Sea or the Nile. The social science-disciplines are also confronted with complex processes of change, particularly as the Beja are becoming involved in "modern sector activities". Through the field works of geographers and anthropologists adaptive strategies can be mapped, both traditional agro-pastoral ones and new ones. Life histories and local conceptualizations of current and former changes are valuable sources through which an understanding of the regional game can be obtained. But the wider contexts of markets and politics are also important as they are affecting the opportunities available for the Beja. And finally, as people in pastoralist production systems as those found in the Red Sea area are constantly sloughed off and have to leave their adaptation, the availability of alternative employment is also an important research issue.

The fact that different disciplines work on different types of data and have different perspectives on development is an important starting point for interdisciplinary discussions within the project. On a general level such debates relate to the classic debates about the relationship between nature and culture. A second challenge is to deal with information from different levels of scale. Processes of change are locality-specific in the sense that they manifest themselves in specific areas and must be studied empirically there. But non-local processes and dynamics also have to be studied. Markets as well as political and administrative systems all provide dynamics that in important ways constrain the local community. They pay taxes, they trade, go on labor migration etc., thereby becoming participants in larger processes of change that are termed commercialization and modernization and which bring about new patterns of differentiation. For an inter-disciplinary programme like RESAP a major challenge is to find ways that may cut across these disciplines, facilitating an understanding of the Red Sea region as a viable concern, and thus improving on the former fragmented and sectoral understanding.

Applied Research and Academic Competence Building

We have said that the aim of RESAP and other SSE-projects is to combine academic competence building in the SSE-countries and Norway with applied research that is relevant for the development issues in the region. Such a combination of aims has for a long time been central in the cooperation between the universities in Bergen and Khartoum. Since the 1960's researchers from both universities have been cooperating in research and education[7]. Within social anthropology common research interests on various development problems, particularly in rural areas, have made a history of staff exchange and of producing students at M.A. and Ph.D. levels possible. From the beginning an underlying idea for this particular cooperation has been to build research competence by tying the research issues to problems that are of relevance to development. Pastoral development, agro-pastoralism, labor migration, desertification as well as problems in the irrigation sector can only serve as key words for such problem areas. An important point is that the research problems were formulated in ways that opened up for cumulative research. Hypotheses about dynamics of different land use

---

7. The cooperation between Bergen and Khartoum goes back to the 1960's. Anthropology, history and archaeology all have long histories of contact. In 1981 a formal agreement was signed broadening the basis for cooperation between the two universities and today there is cooperation going on within all faculties.

systems within larger regional contexts opened up for a number of studies that looked at such processes from different angles involving different disciplines. Such studies gave rise to work of theoretical as well as of empirical value. The applied pay off from this is the increased understanding of the problem areas mentioned above. The same insights could also be made use of in different empirical contexts, and the competence developed within this cooperation has been made use of in many applied contexts.

This cooperation has all the time been based on the interests and initiatives of researchers, and it has expanded slowly, when new common grounds has been established. Researchers have visited each other and have developed an understanding of possibilities and constraints with the other party. The organizational framework for cooperation was also chosen on the basis of what was most suitable at any one point in time. Thus the Department of Anthropology, the Social and Economic Research Council as well as the Development Studies and Research Center have at different times been the main cooperating body.

A basic factor for the continuation of this process has been the constant support from university authorities at various levels. A basis for expansion was also provided by the support rendered by NORAD and later MDC. Financing fieldwork, visits, and scholarships for students these development institutions have been important actors in the development of the Bergen-Khartoum link. Such finance has also been the basis for the so-called "Savanna Project" through which the anthropological cooperation expanded and in which other social science disciplines also participated. To-day, MDC/NORAD finances RESAP as well as other programs involving Bergen and Khartoum.   In this context of cooperation between the universities in Bergen and Khartoum, RESAP is the most recent effort to continue work focussed on land use and regional systems, on society and environment. The aim is to continue and enlarge earlier efforts within this cooperation, but also to realize the ambitious aims for the research component of the SSE-program. To what extent we shall succeed remains to be seen, but without the lessons and guidelines offered by the earlier experiences, and without the relationships created through that process, our chances of success would definitely have been smaller.

The Vegetation of Semi-Arid Areas
A Potential for Economic Development?

Vagn Alstrup

Introduction

Semi-arid areas are typically characterized by a low and often unstable precipitation of between 150 and 400 mm per year, but this can vary according to local conditions. The rain is usually concentrated in one or two annual rain-seasons, interrupted by dry seasons of varying length, it often falls in local showers and is not evenly distributed. The natural vegetation is adapted to these conditions, in that it can survive prolonged drought periods, and develops rapidly after rain.

Agriculture is risky in semi-arid areas, as the harvest may fail due to drought. The conditions are best in hilly areas, which have a higher precipitation than the surrounding areas, and at the foot of hills, which receive run-off moisture. Permanent settlements are therefore often found at the foot of hills. Forest vegetation on mountain slopes retains rain water, delaying the run-off, thereby stabilizing the supply of water to the downhill areas. The clearing of sloping terrain for agriculture, increases the run-off moisture, and often results in a prolongation of the dry season and soil erosion, thereby disturbing the ecosystem.

Rain-fed agriculture in semi-arid areas occurs normally in the form of subsistence farming, practiced as shifting cultivation with simple tools. The settlements are periodic, and fires are used to create new fields, when the productivity goes down in the old fields. Subsistence farming involves a large amount of man-power, though it only supplies a family with grain and vegetables, i.e. it does not produce much surplus for sale. Subsistence farming can be improved by adapting to agroforestry, a system in which some trees are left, which will protect the soil from erosion and improve the local climate. Annual crops are then cultivated between and under the trees, giving a higher yield, and, in addition some wood which can be sold in times of harvest failure, thus helping to stabilize the owners economy in such periods.

In contrast, cash farming, which produces crops to be sold, is practiced with machinery, fertilizers etc. and needs large financial investments. The losses can therefore be considerable in cases of harvest failure, and consequently cash farming should only be practiced in areas of sufficient, and stable precipitation. Cash farming can also be practiced as agroforestry, in which the trees are arranged in rows, but in most cases it is not possible to transform subsistence farming into cash-farming in semi-arid areas, as the production costs will be too high for its products to compete on the world market.

As a consequence of the rapidly growing population, the numbers of subsistence-farming families are increasing beyond the optimum, leading to a gradual impoverishment of the natural vegetation. This is because the periods of lying fallow become too short to allow the vegetation to redevelop in the formerly cultivated fields, which in turn leads to the cultivation of still more marginal areas and hill slopes. This trend can only be changed by an economic development, which will give people better opportunities than subsistence-farming does.

Many resources of wood, meat and other products are lost in these areas due to inadequate handling, lack of storage capacity, inefficient transport systems etc.. Therefore an economically viable production process must be developed and applied via the introduction of new methods and the improvement of the existing infrastructures. However, economic development of these areas is only possible through industrialization, which apart from in the mining industry must be based on the utilization of plant production either directly, or via animal production.

The Natural Vegetation in Semi-Arid Areas

The natural vegetation normally consists of a dense growth of different shrubs and small trees, forming an upper level of up to 5-6 m height, and a lower level of small shrubs, herbs and grasses. In the case of dense thickets, the lower layer is only slightly developed, as it is dependant on access to sufficient light. The herbs and grasses are only present in the wet seasons, and the shrubs and trees are mostly deciduous. Since this vegetation system is conservative in its soil and water consumption (because the precipitation is withheld thereby reducing surface run-off), the same pattern should be maintained in any economic exploitation of it, since plant production in such a system can be relatively high.

Such dense thickets are found in the eastern Kitui district of Kenya. They are mainly composed of species of Acacia, Combretum, Terminalia, Grewia, Cordia, Maerua, etc. Many of the trees and shrubs are spiny, and poisonous species are frequent in both layers and in the shrub layer, which is a consequence of heavy browsing, which reduces the frequency of edible species. Soil erosion is serious in some places, especially in areas cleared for charcoal burning, along roads on sloping terrain and on hills cleared for cultivation.

In coastal Somalia, south of Mogadishu, a 10-20 km broad dune zone is found along the coast. The conditions for tree growth are quite favorable on the dunes, as sand withholds rain water better than clay. The natural vegetation is luxurious and rich in species, of which many are endemic. However, in some places - where the trees have been felled for use as firewood and charcoal in the towns, and for heating of coral limestone in lime-kilns - the dunes have started to drift.

Along the Shebelle river, in southern Somalia, clay deposits form a broad zone. Close to the river the conditions are good for plant growth due to frequent flooding and high ground water level, but most of the riverine forest has been felled to allow for irrigated agriculture. At greater distances, where flooding are rare and ground water not available, the conditions are very harsh. Some hardy species, such as Thespesia danis, grow well and improve the soil, but if the trees are cut, the soil structure is disturbed and a vertisol is formed, on which successful germination of tree seeds is almost impossible. The areas are then left as degraded bushland or grassland. The soil suffers from a water deficit and is of low productivity, except in the rare cases of extraordinarily heavy rainfall.

The typical savanna vegetation is a plain of grass and herbs with scattered trees and/or groups of trees. Its origin is disputed, but it is likely that the reduced tree cover is a result of either deliberate burning, or effective browsing (eating of the leaves of woody species) which over a longer period of time has prevented tree seedlings from surviving. Grass production on the other hand is rather high, and the resulting dense grass cover also prevents successful germination of tree seeds. The microclimate is harsh, and thus savannas are not well suited as grazing areas for livestock, or for agriculture. The most promising vegetation composition lies somewhere between the savanna type and the bush thicket, both of which are rather hostile environments.

Animal Production

Pastoralism is doubtless already the most important economic use of the vegetation, and can presumably be effectively developed with a considerable rise in meat production as a consequence. Cattle, goats, and in dryer areas also camels, are the most important livestock. Wildlife is also an important meat source, and a combination of different plant-eaters will probably give the highest yield, since different animals prefer to eat different plant species.

Transportation of fresh milk is expensive and difficult, thus milk production is only economical for local consumption, and that of nearby towns. Conservation of meat is also difficult in the hot climate, and animals are only slaughtered for a day's consumption at a time. In the case of long drought periods, meat prices drop rapidly to a few percent of the normal price, and many animals die from thirst and starvation. Leather is a rather expensive material and easy to preserve and dispatch, but much skin is lost due to the lack of a modern tanning industry.

Great improvements can certainly be obtained through better breeding systems and veterinary control. Improvements can also be achieved through better management of the plant resources. The best areas for grazing and browsing are those with relatively few high trees, which give shelter and a little shade, but permit the development of a luxurious ground cover of herbs, grasses and low shrubs within the reach of animals. Such a relatively open two-layered system will give good micro-climatic conditions, soil and water will be retained, and only a minor part of the plant production will be transformed into wood.

Concentrated grazing, and grazing with only one or two animal species, may lead to a deterioration of the grasslands, as inedible species are left untouched, thereby increasing in number and eventually leading to bush encroachment. This situation is counteracted by the presence of a diversity of plant-eaters. It may, however, still be necessary to have extra regulation of the vegetation, for instance the destruction of unwanted vegetation and reduction in numbers of trees etc.

The number of animals should preferably be greatly reduced in the dry seasons, only breeding animals should be left, as the greatest damage to the vegetation takes place in the dry seasons, when fodder supply is limited and the animals may start to eat bark. Furthermore, animals loose weight during the dry season. An optimal development of the meat production sector is therefore also dependant on a large slaughter capacity, cold stores and a meat conservation industry. As many jobs are involved in the creation and running of a meat industry, a greatly increased storage capacity will reduce the risks of hunger disasters. Furthermore, meat production is cheap in such a system and as a result the products will be competitive on the world market providing the foundation for economic development.

Tree Products

The natural vegetation of shrubs and low trees is not suitable for timber production because the boles are too short. Although foreign tree species may grow higher, timber production will still be restricted to the wetter areas, such as hilly slopes, where the growth conditions are more favorable. If clear-cutting is avoided in these areas there will be a low risk of soil erosion. Timber can also be produced along rivers and wadis, but irrigated agriculture is often preferred in such places. In drier areas, wood production can be combined with grazing and browsing.

The uses of wood will mainly be restricted to the manufacture of tools, handicrafts, charcoal, firewood and local building purposes. Some of these products can be sold.

Trees are often cut in a height of about one meter above the ground, because the tools used for felling are simple. As the trees are about 5 m high, more than 20% of the wood is wasted in this way, and the introduction of better tools will be beneficial.

Apart from wood, many trees give other products such as fruit, honey, oils, gums etc. Many of the fruits are edible, and can give an important addition to the daily diet, as they are rich in vitamins(some of them also in proteins). Tree fruit production is less risky than agriculture as the trees normally survive serious droughts, and initial investment is not lost in the failure of a season's crop. There seems to be a potential for the production of saleable surplus tree fruit, especially of species like nuts which can be stored for longer periods. In contrast, since the transport of fresh fruit is expensive, often exceeding the selling price where production takes place far away from the nearest town, increased production of this crop is not as economically desirable. Oils and gums from several trees are already important products, as are the leaves of different species for medicine. However, through research on the chemistry of tree-saps, and pharmaceutical analyses of medicinal plants, many substances will undoubtedly be found, which can be utilized in the chemical, medical and food industries.

The transformation of the natural (and in most places degraded vegetation of economically rather unimportant species) into a highly managed system of productive species, is a tremendous task, inasmuch as it is not possible to just replace the present vegetation with a new system. To start successfully, young trees will need the favorable microclimate found under higher trees. The improvement in species composition must be gradual and is labor intensive, but the economic investment is relatively small, even though reduction in number of animals, to avoid overgrazing, may yet be looked upon as a great investment by poor families, and in contrast with traditional farming methods.

Social Organization

In arid areas and in the driest parts of semi-arid areas, nomadism is a common lifestyle, as people and animals have to migrate according to the localization of the sporadic precipitation. Furthermore, people also move around in areas of shifting cultivation. In such systems, access to unoccupied land must be available to the migrant population, and problems arise in relation to the proper management of the land. As the population and livestock numbers increase, the exploitation of vegetation may exceed its production, while its natural regeneration in a fallow period,(ie. between periods of settlement), may be too slow for the areas to regain their natural productivity. It is then necessary to plant trees and to take other measures to prevent a breakdown of the system, and this in turn calls for diverse forms of investment. These are usually linked with ownership, or similar exclusive legal land right, as no private individual wishes to invest in public areas. Ownership of land can probably only be maintained if the owners or their representatives are permanently present. Through legislation, people can be given the duty to plant new trees to ensure the replacement of those cut, however such a legislation is difficult to control in a system of migrating people.

The Impact on Global Climate

It is postulated, that an increase in atmospheric content of carbon dioxide (CO_2) will lead to a temperature increase on a global scale (the greenhouse effect). This in turn, it is believed, will raise the water level of the seas, through water expansion and a simultaneous reduction of the polar ice cap volumes. The rise of the atmospheric CO_2 content is partly a consequence of the burning of fossil fuels, but the reduction of plant cover, seen in many tropical areas, is also of great importance (i.e. the change of wooded areas into grassland, releases the CO_2 bound in the wood into the atmosphere).

The consequences for Africa of an increase in the atmospheric

CO_2-concentration are disputed. Some believe that a higher temperature will lead to a drier climate thereby constituting an environmental risk for the semi-arid areas; while others are of the opposite opinion, namely that a higher temperature will raise the evaporation from the seas, resulting in higher air humidity and precipitation, especially of benefit to Africa. If the last hypothesis is correct, investment in the management of the vegetation of semi-arid areas will be highly beneficial, as it will lead to a higher plant production. If the opposite hypothesis is right, then management is even more necessary, as a dense plant cover will counteract the rise in atmospheric CO_2, and reduce the consequences of the greenhouse effect. In both cases there is good reason to take care of the vegetation.

Improved Utilization of Multipurpose Tree Legumes Through a Better Understanding of their Chemistry

The Acacia Species: A Case Study. An introduction to selected topics concerning minor forestry products, toxic and antinutritional constituents.

Leon Brimer

Abstract

The following article discusses the present body of knowledge regarding the chemical identity of the toxic and antinutritional constituents of trees and shrubs in the genus *Acacia*, as well as that of the minor forestry products thereof. The discussion focuses upon a number of selected topics. Firstly, those related to the following chemical groups *a* alkaloids, *b* cyanogenic constituents, *c* non-protein amino acids and *d* proteinase inhibitors (trypsin inhibitors - "TI"'s), and secondly, two more functionally defined topics, namely *e* Acacia material employed in (ethno)medicinal preparations and acacia constituents with *f* molluscicidal effect.

The paper concludes with the contention that a monograph covering the topic "in toto" is urgently needed. Such a monograph would be a valuable tool in the selection of the species best suited for different multipurpose usage.

Key Words

Acacia, alkaloid, amino acid, antinutritional, cyanogenic, ethnomedicin, ethnopharmacology, *Leguminosae*, minor forestry product, molluscicide, non-protein amino acid, proteinase inhibitor, schistosomiasis, toxicity.

Leguminous plants constitute a large, natural, and widely prolific group, comprised of about 650 genera and 18,000 species (Anon. 1979), most of which are trees or shrubs - although both herbaceous species and lianas are represented, also. Within the legumes, the genus *Acacia* forms a huge group of woody plants - trees and shrubs - distributed throughout tropical and subtropical regions (Vassal 1972). A remarkable feature of many acacias is, that they tolerate warm and dry conditions, although some, especially some of the so-called swollen-thorn acacias of Central America, may inhabit wetter sites than most other acacias (Janzen 1974).

Among trees investigated and recommended for (re-)afforestation (Lundgren & Nair 1985; Turnbull 1987), mine-revegetation (Langkamp 1987) and the prevention of hillside erosion (Sheppard & Bulloch 1986) in dry and semi-dry climates, several *Acacia* species play an important rôle. In addition to their possession of a general ecological plasticity in dry environments (Nongonierma 1977) and a competence for nitrogen fixation (soil enrichment), this is due to the existence of several species suitable for sand dune stabilization, fuel - and hardwood production (Shakla et al. 1987; Turnbull 1986; Anon. 1980/1981.) and as browse species (Turnbull 1987; Skerman 1977). Furthermore many *Acacia* species are highly valued for their production of e.g. polysaccharide gums (Seif El Din 1975; Freudenberger 1988), tannins (Raymond 1951; Indikumana, Danilkovich 1988), flavour/perfume materials (Peyron 1972; Naves 1974; Führer 1971) and honey (Crane 1985), in sum, the various minor forestry products described below.

Although many acacia species are valuable resources (as indicated above), others contain toxic or antinutritional metabolites in concentrations which make them a definite threat to livestock, a general feature they share with their relatives in the Leguminosae family (Summerfield & Roberts 1985; Huisman, van der Poel, Liener, 1989).

Recently, both the general biology and the special germination problems seen in more than a hundred hardcoated species used in mine-revegetation, have been presented in book form (Langkamp 1987; New 1984; Doran et al. 1983,). Likewise, the first reviews concerning the biotechnological manipulation of acacias (i.e. "in vitro" cloning via the induction of shoot-tip callus followed by regeneration) are now emerging (Turnbull 1987; Skolmen 1986). Furthermore, new checklists and related (electronic) worldwide taxonomic databases are on their way (Lock 1989; Bisby 1988).

However as yet, no monograph exists which tries to incorporate our total, and quite broad, body of knowledge concerning the chemistry of different groups (subgenera/sections) and single species of acacia. By "The chemistry" is meant "the present state of our knowledge about the (mostly secondary [1]) constituents found in the different plant organs, their biosynthesis and regulation, their biological/physiological effects on other organisms, and their (technical) usage/value - if any".

The objective of this article is to demonstrate the need for such a monograph as a tool, both for the scientist and for those making decisions about future forestry/agroforestry projects. In other words, to stimulate an interest in and establish a basis for a broader survey of the possibilities for food/fodder and non-food applications of *Acacia* products. That such a project must be undertaken in an integrated manner, is clearly evident, and further exemplified by Rexen and Munck in their outstanding report concerning the industrial use of cereal crops (1984).

No matter what the function of a given compound in an organism/group of organisms may be, the mere presence of this compound may be of importance - not only due to its effects/potential usage, but also as a tool in the biological systematics. This is true as long as the constituent is found to be restricted to one or more well defined group(s), taxon/taxa, of organisms. Several such compounds are found within the genus *Acacia*.

In the ensuing discussion examples will be given of constituents found in acacias and biological effects described for e.g. extracts of acacia material. The examples, which have been selected from the scientific literature, represent only a tiny fraction of the scientifically based knowledge on the minor forestry products extracted from, and toxic constituents in *Acacia* species.

Our examples have been selected from the underscored items in the list below (showing the most prominent topics to be considered when dealing with acacias):

- *alkaloids*
- *cyanogenic glycosides*
- flavonoids (including tannins)
- honey
- lectins (phytohaemagglutinins)
- *molluscicidal compounds*
- *non-protein amino acids*
- organic flour compounds
- pest controling properties
- polysaccharides (gums)
- *proteinase inhibitors (trypsin inhib.)*
- sensitizing agents
- terpenoids (e.g. perfume remedies and saponins)
- tumor promoters or inducers

In addition we have included information on *Acacia* material reported to be incorporated in "traditional" medical preparations.

Alkaloids

The vast group of secondary constituents most often placed under the heading of *alkaloids*, ranges from simple amines (corresponding to common protein amino acids) to highly complicated structures of mixed biogenetic origin. Alkaloids are well represented within the family *Leguminosae*. Thus well over 350 different alkaloids have been described from the subfamily *Papilionoideae*. Turning to the remaining subfamilies, i.e. *Caesalpinioideae* and *Mimosoideae* (including *Acacia*) both the number of structures and their complexity become more limited. With regard to chemotaxonomy within the genus *Acacia*, mainly the discontinous distribution of **phenethylamine** and **tyramin** (together with their derivatives) have been of value. Thus, both the occurrence/non-occurrence of **phenethylamine** and of **N,N-dimethyltyramin** (**Hordenin**) contribute significantly to the formation of distinct plant groups (taxa) in a multicharacter data analysis of Australian *Acacias* (Pettigrew & Watson 1975).

From the toxicological point of view at least one syndrome of poisoning (intoxication) is well known. Thus, the so called "Quajillo wobbles" = "Limberleg" has been described as the result of consuming great amounts of foliage from *Acacia berlandieri* which contains the abovementioned compounds (Price & Hardy 1953; Camp & Norvell 1966; Evans et al. 1979). It should be further noted, that **N,N-dimethyltryptamin** (resembling the neurotransmitter serotonin) has been reported as present in three species in Northern Sudan (Khalil & Elkheir 1975). This compound is well known for its ability to induce the neurological alkaloid toxicosis "Phalaris staggers" in sheep and cattle (Hartley 1978). Recently, toxicosis has even been identified in cattle grazing on *Phalaris* species selected for low alkaloid content (Nicholson et al. 1989; Lean et al. 1989) - pointing to the importance of this compound from a veterinary point of view.

Note also that selective breeding with a view to reducing the level of certain alkaloids, has proved successful in several legume species, notably *Lupinus* spp. (Huisman et al. 1989).

Cyanogenic Glycosides

Rarely is the nitrile group found in natural compounds, except for its occurrence in cyanogenic compounds. Thus, only a few other compound classes are known to incorporate this chemical entity to any significant level. As examples one could mention certain non-protein amino acids (as e.g. beta-cyano-L-alanine) and the relatively small group of so called 2-cyclohexyl-ethanenitriles. Furthermore, some of the products of rearrangements resulting from the hydrolysis of glucosinolates also contain the nitrile group.

The most salient feature of the cyanogenic compounds, viz. their ability to release hydrogen cyanide (prussic acid) as a product of (most often enzymatic) disintegration, is also the main reason for their classification as toxins. The principal target organ of cyanide is probably the central nervous system (CNS), resulting in either *a* acute (and possibly lethal) intoxication, or *b* chronic low-level exposure (leading to pathological lesions in the CNS) which manifests itself as visual dysfunctions and neuropathic ataxia (Way et al. 1988). Very recent findings support this view. Aitken et al. (1989) for instance, found a decreased synaptic transmission in the hippocampal region ("in vitro") at concentrations of about 10 micromolar of HCN, while Yamamoto has shown that, the loss of consciousness in cases of HCN intoxication (mice) is correlated with the resulting formation of hyperammonemia and change in brain neutral amino acid levels (Yamamoto 1989). While acute intoxication is known for both animals and humans, the chronic syndromes have only been described for humans. A well known example of chronic cyanide intoxication syndrome is the so-called tropical ataxic neuropathia, "resulting" from the daily intake of subacute doses of cyanide/cyanogenic compounds (glycosides) through food consumption (mainly Cassava = *Manihot esculenta* (Rosling 1986, 1988).

The pure cyanogenic compound stored in the organs of a higher plant will be found in the form of either a glycoside, a lipid, a cyanohydrin or a 2,3-epoxynitrile In cyanogenic acacias the main form consists of glycosides, although it seems that small amounts of cyanohydrins may also be present (Conn 1985; Brimer 1988). Present knowledge concerning cyanogenic constituents and cyanogenesis in the genus *Acacia* is derived from very detailed studies of the constituents in single species or varieties. It also stems from extended screening programs, concerning the ability to liberate HCN with (or without) exogenously added hydrolytic enzymes. Perhaps one of the most interesting details is the fact that few of the *Acacia* species with cyanogenic constituents seem to possess hydrolytic enzymes (glycosidases) with preference for these substrates (Conn et al. 1985). This is in strong contrast to the present knowledge concerning most other taxa with cyanogenic metabolites and species which contain the biogenetically related glucosinolates - all of which possess the degradating enzyme(s) myrosinase(s) (Rodman 1978).

From the onset of research into the cyanogenic constituents of *Acacia* species (i.e. prior to the definition of the rules of chemotaxonomy), chemists and botanists witnessed an instance of chemosystematic classification within a particular genus, namely, the genus *Acacia*. Thus, even the two earliest analyses of the structures of the cyanogenic glycosides to be found in acacias (Rimington 1935; Finnemore & Gledhill 1928) hinted at the main difference and point of interest today. Accordingly, Nartey (1978) noted that "- it appears that the *Acacia* is the only genus, in which cyanophoric species occur, which contain cyanogenic glycosides, with both aromatic and aliphatic aglucones". A few years later Seigler and Conn in the overview "Cyanogenesis and systematics of the genus *Acacia*" stated that "Plants from series *Gummiferae* (syn. subsp. *Acacia*) synthesize cyanogenic glycosides which are derived from the aliphatic aminoacids. On the other hand, all of the Australian species studied synthesize compounds from the aromatic aminoacid Phenylalanine" (Seigler & Conn 1982).

Today some 70 species worldwide are known to contain cyanogenic compounds. 45 of these are placed in subgenus *Phyllodineae* (syn. *Heterophyllum,* of australian origin), 20 in subgenus *Acacia* (3 of these typical ant-acacias (Seigler & Ebinger 1987)) and 5 in subgenus *Aculeiferum* (Seigler & Conn 1982; Maslin et al. 1988; Brimer 1986). The concentrations in the foliage being in the interval from 1 to 90 micromolar per gram and up to about 90 (Maslin et al. 1988).

The differences observed in the types of cyanogenic glycosides synthesized (ref. discussion above) remain valid, insofar as all identifications (but one) of constituents in species from *Phyllodineae* have shown mixtures of sambunigrin with prunasin (aromatic aglycones), while species from the pantropic subgenus *Acacia* so far have been characterized by aliphatic aglycones (Seigler & Conn 1982; Seigler & Ebinger 1987; Maslin et al. 1988; Brimer 1986). The few identifications in subgenus *Aculeiferum* showed the presence of mixtures of sambunigrin and prunasin, as seen in the Australian *Phyllodineae* (Conn et al. 1989).

As for the toxicity of ingested cyanogenic glycosides feeding studies have been carried out with cattle, sheep, pigs, dogs, rats and hens/ chickens. Furthermore, clinical observations and results, from a number of laboratory experiments, have been reported for humans following oral, rectal and parenteral administration of amygdalin. Experiments with ruminants have shown that the cyanogenic glucosides linamarin and lotaustralin are readily and quickly hydrolyzed in the rumen of sheep (Coop & Blackley 1949; Blackley & Coop 1949; Coop & Blackley 1950), and that this may lead to acute toxic reactions and death. "In vitro" experiments with rumen liquor from a straw-fed fistulated cow have further indicated that detoxification of HCN formed in the rumen is quite unlikely to occur (Ngarmsak 1977/78). Both of these results thus support the "general view" that ruminant animals are more susceptible to acute HCN poisoning than are monogastric animals (Kingsbury 1964). Most (if not all) reports on acute intoxication due to the intake of cyanogenic glycosides, concern sheep (*Acacia* species) or cattle (*Sorghum* species and *Triglochin maritima* = arrow grass) (Conn 1978; Majak et al. 1980). However, these are also grazing and browsing animals. In contrast, no experiment so far has detected signs of chronic toxicity or affected weight gain in ruminants, as concluded by Hill (1973). However, such experiments have been few in number and limited in their analysis of the effect of different concentrations of the cyanogenic glycoside as well as that of the total composition of the diets (Hill 1973; Ngarmsak 1977/78).

This was more or less the situation (concerning ruminants) when Majak and Cheng (Majak & Cheng 1984) published their paper on the release of HCN (from *Amelanchier alnifolia* - a saskatoon serviceberry, belonging to *Rosaceae* and containing prunasin) and the detoxification of the HCN formed when fed to rumen fistulated Hereford heifers on *different* diets. The investigation ,which also included "in vitro" experiments, concluded that: (a) all heifers and rumen inocula could hydrolyze prunasin and amygdalin, although the rate of HCN formation showed great variations, (b) this formation was independent of the composition of the diet (5 types were administered), (c) rumen bacteria were responsible for the hydrolysis (several species were shown to be active), (d) thiocyanate could be formed from the HCN by the rumen bacteria, however, this mechanism did not play any major rôle concerning HCN disappearance. No signs of acute toxicity were observed using a single administration of berries corresponding to 2.5 mg HCN/kg.

Chickens, on the other hand, seem to be quite susceptible, since both linseed meal, cassava meal, and cassava leaf meal may cause retarded growth and increased mortality when added to the diet. All three products contain parent cyanogenic glycosides or products thereof, gained through hydrolysis. Growth has been improved by increasing the methionine content of the fodder or by soaking it prior to use (Hill 1973). It also seems that fully grown chickens possess higher tolerance levels.

Apparently swine tolerate linamarin, at least up to a certain level. This is indicated by several feeding experiments in which the quite extensive use of cassava in pig diets has, in general produced acceptable weight gains, without signs of toxicity (Hill 1973; Best 1978).

Turning to the acacias, a recent discussion on the toxicity potential of the australian acacias found to contain cyanogenic glycosides (Maslin et al. 1987), may give us some guidelines. Thus, a conclusion based on this paper and on the fact that only few African acacias (e.g. *Acacia sieberana* var. *woodii*, *A. hebeclada* and *A. erioloba* - all from Southern Africa) have been described as containing cyanogenic glycosides at concentrations that may kill or are known to kill sheep (Rimington 1935), must be: that only a limited number of species possess the ability to produce cyanogenic glycosides above a level of about 20 mg of HCN per 100 g fresh weight - the concentration at which a possible risk may be allowed for.

However, polymorphism for both cyanogen and hydrolytic enzymes is well described within *Leguminosae*, notably the species *Trifolium repens* and *Lotus corniculatus*. Polymorphisms have also been observed in the acacias (Conn et al. 1985; Maslin et al. 1988; Maslin et al. 1987), as have great variations in the concentration of cyanogenic glycosides in species synthesizing these (Maslin et al. 1987). Furthermore, there have been indications that hybrids originating from otherwise typical specimens may in some cases have led to cyanophoric qualities in otherwise non-cyanogenic species (Seigler & Ebinger 1988; Brimer et al. 1987).

To the forester, this means that the material should be checked for the properties mentioned. Methods for the quantitative analysis of cyanogenic glycosides and of testing for the presence of enzymes are available for both laboratory (51) and field contexts (Brimer 1988; Brimer & Mølgaard 1986).

Non-Protein Amino Acids

A group of secondary constituents which were recognized quite early on for their significance in the chemotaxonomy of the genus *Acacia*, as well as in the systematics of the entire group of leguminous plants, are the non-protein amino acids. These substances also constitute an important group in relation to toxicity. Humans, grazing and browsing domestic animals, and unwanted plant predators (e.g. insects), have been described as targets for such poisonings. Of the about 250 different non-protein amino acids known to occur in nature, more than 80 have been found in one or more species from the family *Leguminosae* - a higher number than in any other plant family (Bell 1981).

Within the subfamily *Mimosoideae* (including the acacias) especially, the distribution patterns for N-acetyldjenkolic acid , S-carboxyethylcysteine, albizzine, diaminopropionic acid (together with its oxalyl derivative ox-dapro) and different alpha-substituted glutamic acids have been of interest to taxonomy. Turning to the genus *Acacia*, its patterns are of similar interest, excluding those of its glutamic acids which together with S-carboxyisopropylcysteine and alpha-amino-beta-acetyla-minopropionic acid play the role of marker compounds. The borderlines for their discontinous distribution correlate perfectly with the systematic groups of Vassal (1972).

Returning to the seven, taxonomically significant acids, we find that two of these are well-known toxins or derivatives of toxins. Thus beta-N-oxalyl-alpha,beta-diaminopropionic acid is the most prominent of the toxins responsible for the feared lathyrus intoxication *human neurolathy-rism* (Misra et al. 1981). Furthermore, N-acetyldjenkolic acid is a derivative of djenkolic acid, well known from *Pithecellobium lobatum* to cause damage of the kidneys, crystallizing there as well as in the urinary tract. What is more important, however, is that the djenkolic acid itself was actually also found in most of these species (Evans et al. 1977).

Unfortunately, Evans et al. (1977) only give the concentrations in a relative score system (+ to +++). According to the scores given for ox-dapro, toxic concentrations might be found in seeds from subgenus *Aculeiferum*.

It must be emphasized that no toxic syndrome (human or veterinary) has been ascribed to the intake of acacia material rich in non-protein amino acids. This is in contrast to the situation of several other legumes, used as fodder or recommended as a source thereof; one example being the problems of *mimosine* intoxications (leucaena toxicosis). This results from the intake of great amounts of *Leucaena leucocephala* by ruminants, if the breakdown product 3-hydroxy-4(1H)-pyridone (3,4-DHP) cannot be further detoxified by the animal. This detoxification is dependent on the presence of 3,4-DHP-degrading rumen bacteria (Hammond et al. 1989). Mimosine has not been detected in *Acacia* species.

Proteinase Inhibitors (Trypsin Inhibitors - "TI")

A group of macromolecular compounds (besides the gums) that have proven valuable in the systematics of genus *Acacia* is that of the so-called proteinase inhibitors. Many of these inhibitors - which are of a protein nature themselves - are quite stable to low pH and peptic digestion, and thus able to reach the small intestine relatively un-scathed. Once in the intestine, they may inhibit the pancreatic pro-teinase of the animal, decreasing the availability of dietary protein for absorption. This is mainly a problem for non ruminant animals. Most of the proteinase inhibitors known have been described from seeds of leguminous plants - the outstanding examples being those of : *Glycine max* (the soybean), *Arachis hypogaea* (the groundnut/peanut) and *Phaseolus lunatus* (the lima bean), from which isolated inhibitors had their amino acid sequences determined in 1970, 1972 and 1974 respectively (Ory 1981).

With regard to the systematics of genus *Acacia*, the basic work was laid down by Weder, who examined not only the activities, but also the number of bands and their migration during electrophoresis - by employing a specific screening for trypsin (TI)- and chymotrypsin inhibitors respectively (Weder 1981; Weder & Murray 1981). The results from these investigations allow the conclusion that the acacias may be distinguished from other genera within the *Leguminosae*, and that a subdivision of the genus based on the electrophoretic pattern do show an acceptable correlation with other classifications.

Isolation and partial structure elucidation (sequencing) of inhibitors from acacias were reported in 1981 (*A. elata*, Kortt & Jermyn 1981) and 1983 (*A. sieberana*, Jourbert 1983) respectively. In both cases, two inhibitors were found. All four resembled the Kunitz - type (high molecular weight) found in soybean, i.e. a MW about 20.000.

It is rather complicated - if not impossible - to predict the possible nutritional significance of a given proteinase inhibitor even under controlled conditions (animal spp., diet, preparation of diet). Thus, even though Rackis and Gumbeman (1981) state "Ever since Osburn and Mendel (1917) observed that soybeans would not support the growth of rats unless the beans were cooked for 3 hours on a steam bath, the nature of the substances that inhibit growth, enlarge the pancreas, and cause other physiological and biochemical effects in various species of animals, has been investigated intensively", the same authors six pages must later conclude that "The practical significance of TI's with respect to human nutrition at the moment is speculative". The expressed uncertainty is due to several factors, all relevant for such a judgement - no matter if it is done for humans or for other monogastric species:

1. In some species (e.g. man) pancreatic feedback inhibition occurs, and since only about 2% of human trypsin is inhibited by an equivalent weight of soy TI, very high levels may be needed for the production of a significant effect.
2. There are great differences in the potency of a given inhibitor towards trypsin and other proteinase from different animal species.

3. Most reported studies have been done on bovine trypsine and chymotrypsine.
4. Different forms of trypsins (e.g. cationic & anionic) may be present in the same species, as seen in man. These will normally exhibit differential affinity towards a certain inhibitor.
5. Symptoms, such as pancreatic hypertrophy have proven reversible - at least in rats.

It is a fact that fodder containing large amounts of raw legume seeds with active TI (together with other antinutritional factors (ANF's) - tannins, lectins etc.) is less successful in promoting the growth of swine and poultry, than is fodder which has been processed in order to deactivate these factors. Deactivation af TI and lectins - both of protein nature - is achieved through different kinds of heat treatment (Huisman et al. 1989).

Our conclusion then must clearly be that proteinase inhibitors may be present in *acacia* seeds (and perhaps even in other material), and that the effect of these - and other ANF's should be evaluated, if projects including the intensive use of chemically uncharacterized *acacia* material is planned. If processing prior to use is planned, this should be assayed for its ability to improve the digestibility/biological value.

Both qualitative and quantitative variations in the content of trypsin inhibitors have been described as heritable factors for certain legume species (Summerfield & Roberts 1985), thus a potential for a reduction of these factors may also exist in certain *acacia* species.

The Incorporation of Acacia Material in Medicinal Preparations

Different organs from several *acacia* species are known to constitute parts of traditional medicinal preparations from different geographical regions. In nearly all cases the active principle(s) is/are unknown, although in a few cases it seems possible to make a qualified guess, based on knowledge about identified constituents, this is primarily true of all cases in which alkaloids or maybe tannins are known to be present.

Nearly all sources dealing with "medical plants" or "traditional medicine" from one of the main areas of distribution of the genus *Acacia*, quote one or more species as being part of the local *materia medica*.

As part of a general study of the chemistry of the acacias, the present author has perused more than 600 books and articles dealing with this subject. The upshot was that well over 50 species are documented to be used for more than 150 indications. Since local ethnomedicine is often reflected in the earlier pharmacopoeias, it seems natural to analyze the incidence of acacia products in early pharmacopoeias. Most of the information needed for such an analysis is found in Imbesi (1964), which lists 15 *Acacia* species as official entries in one or more pharmacopoeias worldwide - mainly as producers of gums and tannins. An analysis of the former, which consists of approximately 50 species, does not, however, show a very high representation of typical gum species or of gum as the principal material used. Of course this discrepancy is due to the fact that most old pharmacopoeias originate in European countries and make use only of gum arabic as an imported technical compound. On the other hand, *A. catechu* is cited nearly everywhere as a medical remedy, but most often with indications clearly pointing to its use as an astringent (i.e. the medical application of this species).

Another, and more interesting, example is that of *A. tortilis* ssp. *raddiana*, which is used as a remedy against asthma in Somalia. This species has been investigated thoroughly by Hagos and Samuelsson, who found two new natural compounds in gum covered bark (both 1,3 diaryl-propan-2-ol derivatives) which they named *quracol* A and B. Both are potent smooth muscle relaxing agents, i.e. compounds which will actually be active against asthma (Hagos et al. 1987). The authors could further show that the active compounds were situated in the gum and not in the bark, correlating with the observation from the traditional healers - that only gum-covered bark is active (Hagos & Samuelsson 1988).

This latter example indicates that further information on medical active compounds could be obtained through investigations of other *Acacia* species. The odds for a positive outcome of an investigation based on ethnopharmacological evidence (past or present) are known to be high (Farnsworth 1984; Bruhn & Helmstedt 1981).

Finally it should be noted that the above-mentioned *A. tortilis* is among those (African multipurpose) species for which the OFI genetics Group has started genetic development and conservation products, resulting in, among other things, the publication of annotated bibliographies on the species in question (Barnes 1988).

Molluscicides

In the combat against human and domestic animal schistosomiasis (bilharziasis) and other parasitic diseases with water related snails as obligate intermediate hosts (e.g. infections with *Fasciola hepatica*, domestic animal liver flukes) molluscicides have played an ever changing yet important rôle (Tameim et al. 1985; Webbe 1987). The result of years of laboratory screening, field testing, and evaluation of actual snail reduction campaigns is, that to-day only one compound, the synthetic **niclosamide** (BayluscideR) is in widespread use (Webbe 1987; Brinkmann et al. 1988). This situation is totally different from that of most other classes of pesticides (Worthing & Walker 1987), and makes no room for individually adapted - chemically based - eradication strategies. Moreover, niclosamide is regarded as a relatively costly compound by most developing countries.

Facing these realities, expert committees advising WHO, as well as other leading scientists, have in the period from 1983 (Heyneman 1984) to 1986 (Anon. 1987) and onwards recommended further research and development in the field of plant derived molluscicides, notably the saponins from *Phytolacca dodecandra* - the "endod" plant (Heyneman 1984; Anon. 1987). As an integrated part of these initiatives, UNDP/World Bank/WHO supported the preparation of a book including a number of reviews concerning known plant molluscicides and molluscicidal plants (Mott 1987).

It may seem striking, that the *Leguminosae* cover 8 out of the 50 pages of a table on "Dicots Tested for Molluscicidal Activity or reported to contain Active Molluscicides" in this book, however in reality this reflects two facts, *firstly* that this category is very big *and secondly* that it includes several species showing promising activity. These species are found scattered in all three subfamilies (a) *Caesalpinioideae*, (b) *Mimosoideae*, and (c) *Papilionoideae*, examples being (a) *Swartzie madagascariensis*, (b) *Tetrapleura tetraptera* and (c) *Sesbania sesban*. In addition, *Triterpene* saponins have been shown to be the active constituents for quite a number of the active legumes.

Of the 24 *Acacia* species tested so far (Anon. 1987; Mott 1987), active constituents have only been isolated (by applying a bioassay guided procedure), from one species: *A. nilotica* (ssp. *tomentosa, astringens* and *nilotica*). The fruit as well as the bark from these proved to contain (-)-epigallocatechin-7-gallate and (-)-epigallocatechin-5,7-digallate as the two main constituents with molluscicidal activity, the latter being the most active (Ayoub 1985). Both compounds were subsequently shown to be present in the bark/pod of 20 other species and subspecies of *Acacia* from Sudan, and the molluscicidal effect of these extracts were furthermore shown to correlate with that of the concentration of polyphenolic substances (tannins) in the extracts (Ayoub & Yankov 1987).

The 2 abovementioned constituents are members of the vast family of flavonoids, substituted flavonoids and condensed flavonoids (proanthocyanidines or condensed tannins). Limited studies of structural activity relationships have (although weakly documented) pointed to the presenceof a number of free hydroxyl (phenolic) groups, together with some other specified structural features, such as the factors that determine this activity (Ayoub 1985; Ayoub & Yankov 1985).

Among the listed plant compounds showing promising toxicity towards water related snails (Mott 1987), none has been fully characterized for their toxicity towards warmblooded animals and humans (Anon. 1987). This holds also true for ecotoxicological aspects if used in open field systems i.e. applied to or brought into the environment (Anon. 1987). Among the handful of plant species for which these kind of investigations have advanced greatest (or are making good progress), we will find two legumes, namely *T. tetraptera* and *Swartsia madagascariensis*. None of the acacias have been tested in any further detail concerning toxicity.

Further research within the field of the molluscicidal properties of *Acacia* species could reveal valuable information. This is perhaps even more true, given that those species which have been shown to be active, are the same as those which are known to produce valuable tannins for the leather industry - lending hope for the multipurpose utilization of these trees. Further screening for molluscicidal activity could also prove interesting, since the majority of species, including the typical australian species, have not yet been tested.

Conclusion

The general consensus concerning the great value of many acacias for different purposes in dry regions, thus points to the need for knowledge regarding all aspects of these plants - including the chemistry of their forestry products and of their toxic and antinutritional constituents. The present overview has shown that although a bulk of information already exists in this area, it needs to be systematized. A monograph achieving this goal will in addition to greatly benefitting afforestation strategies, certainly also point to the topics that are still in need of investigation. Thus, although many species have been investigated in relation to the antifeedant effects of their constituents (ref. above), many still await similar treatment. As an example one could take *Acacia mellifera*. This plant was found to show high values for "in vitro" digestibility and crude protein . Nonetheless this plant was observed not to be browsed to any significant extent (Skarpe & Bergström 1986). Nothing is known about antifeeding components in this plant.

86

The interest expressed worldwide, in the *Acacia* species, may best be exemplified by the fact, that the authorities on the French island of Corsica established an experimental plantation in 1982 including 2,200 plants representing 19 species, with the aim of investigating their multipurpose utilization (Vassal 1983).

Acknowledgements

The author is indebted to Per Anker Jensen, Department of Computational linguistics, Copenhagen Business School, for valuable discussions and language revision.

References

Aitken, P.G. & Braitman, D.J. 1989. The effects of cyanide on neural and synaptic function in hippocampal slices. Neurotoxicol. 10:239 - 248.

Anderson, D.M.W. 1986. The future of gum arabic as an article of international trade. Bull. IGSM 14:65 - 67.

Anderson, D.M.W. 1986. The amino acid components of some commercial gums. In: Phillips, G.O., Wedlock, D.J. & Williams, P.A. (Eds) Gums and stabilizers for the food Industry. Elsevier Applied Sciences Publishers, London, p 79 - 86.

Anderson, D.M.W. 1988. Exudate and other gums as forms of soluble dietary fibre. In: Walker, R. & Quattrucci, E. (Eds) Nutritional and toxicological aspects of food processing. Taylor & Francis, London, p 257 - 273.

Anon. 1979. Tropical Legumes; Resources for the Future. Report of an Ad Hoc Panel. National Academy of Science, Washington D.C..

Anon. 1980/1981. Firewood crops: shrub and tree species for enegy production: report of an ad hoc panel of the Advisory Commitee on Technology innovation. 1'st/2'end print. National Academy of Sciences, Washington, D.C..

Anon. 1987. Appendix II, Endod Toxicology,Report of the expert Group Meeting, United Nations, New York, 27 - 28 February 1986. In: Makhubu, L., Lemma, A. & Heyneman, D. (Eds) Endod II (Phytolacca dodecandra). Council of Int. Aff., New York, p 131 - 148.

Ayoub, S.M.H. 1985. Flavanol molluscicides from the Sudan Acacias. Int. J. Crude. Drug Res. 23:87 - 90.

Ayoub, S.M.H. & Yankov, L.K. 1985. On the molluscicidal activity of the plant phenolics. Fitoterapia 56:225 - 226.

Ayoub, S.M.H. & Yankov, L.K. 1987. Molluscicidal Properties of the Sudan Acacias. Fitoterapia 58:363 - 366.

Barnes, R.D. 1988. Tropical Forest Genetics at the Oxford Forestry Institute. Commonw. For. Rev. 67:231 - 241.

Bell, E.A. 1981. Non-protein Amino Acids in the Leguminosae. In: Polhill,

R.M. & Raven, P.H. (Eds) Advances in Legume Systematics, part 2. Royal Botanic Gardens, Kew, p 489 - 499.

Best, R. 1978. Cassava processing for animal feed. In: Weber, E.J., Cock, J.H. & Chouinard, A. (Eds) Cassava harvesting and processing. International Development Research entre, Ottawa (IDRC - 114e), p 12 - 20.

Bisby, F.A. 1988. International Legume Database & Information Service 2nd Workshop. Bull. I.G.S.M. 16:6 - 10.

Blackley, R.L. & Coop, I.E. 1949. The metabolism and toxicity of cyanides and cyanogenetic glycosides in sheep. 2. Detoxification of hydrocyanic acid.N.Z.J. Sci. Technol. (A) 31:1 - 16.

Brimer, L. 1986. Cyanide and cyanogenic compounds. The occurrence in material of biological origin and the analysis - with special respect to cyanogenic constituents in *Acacia* species. Dissertation Royal Danish School of Pharmacy, Copenhagen.

Brimer, L. 1988. Determination of cyanide and cyanogenic compounds in biological systems. In: Cyanide compounds in Biology. Wiley, Chichester (Ciba Foundation Symposium 140), p 177 - 200.

Brimer, L. & Mølgaard, P. 1986. Simple densitometric method for estimation of cyanogenic glycosides and cyanohydrins under field conditions. Biochem. Syst. Ecol. 14:97 - 103.

Brimer, L., Ebinger, J.E., Seigler, D.S. & Vassal, J. 1987. Cyanogenesis in *Acacia cochliacantha* (Fabaceae, Mimosoideae). Bull. IGSM 15:88 - 99.

Brinkmann, V.K., Werler, C., Traore, M. & Korte, R. 1988. The National Schistosomiasis Control Programme in Mali, Objectives, Organisation, Results. Trop. Med. Parasitol. 39:157 -

Bruhn, J.G. & Helmstedt, B. 1981. Ethnopharmacology: Objectives, Principles and Perspectives. In:Beal, J.L. & Reinhard, E. (Eds) Natural Products as Medicinal Agents. Hippokrates, Stuttgart.

Camp, B.J. & Norvell, M.J. 1966. The Phenylethylamine Alkaloids of Native Range Plants. Economic Bot. 20:274 - 278.

Conn, E.E. 1978. Cyanogenesis, the production of hydrogen cyanide by plants. In: Keeler, R.F., van Kampen, K.R. & James, L.F. (Eds) Effects of poisonous plants on livestock. Academic Press, New York, p 301 - 310.

Conn, E.E., Maslin, B.R., Carry, S. & Conn, M.E. 1985. Cyanogensis in Australian species of *Acacia*. Survey of Herbarium specimens and Living Plants. Western Austr. Herb. Res. Notes. No. 10, 1 - 60.

Conn, E.E., Seigler, D.S., Maslin, B.R. & Dunn, J. 1989. Cyanogenesis in *Acacia* Subgenus Aculeiferum. Phytochemistry 28:817 - 820.

Coop, I.E. & Blackley, R.L. 1949. The metabolism and toxicity of cyanides and cyanogenetic glycosides in sheep. 1. Activity in the rumen. N.Z.J. Sci. Technol. (A) 30:277 - 291.

Coop, I.E. & Blackley, R.L. 1950. The metabolism and toxicity of cyanides and cyanogenetic glycosides in sheep. 3. The toxicity of cyanides and cyanogeneticc glycosides. N.Z.J. Sci. Technol. (A) 31:44 - 58.

Crane, E. 1985. Bees and honey in the exploitation of arid land resources. In: Wickens, G.E., Goodin, J.R. & Field, D.V. (Eds) Plants for Arid Lands, Unwin Hyman, London, p 163 - 175.

Doran, J.C., Turnbull, J.W., Boland, D.J. & Gunn, B.V. 1983. Handbook on seeds of dry-zone Acacias: a guide for collecting extracting, cleaning, and storing the seed and for treatment to promote germination of dry-zone acacias. FAO, Rome.

Evans, C.S., Quershi, M.Y. & Bell, E.A. 1977. Free Amino Acids in the Seeds of *Acacia* Species. Phytochemistry 16:565 - 570.

Evans, C.S., Bell, E.A. & Johnson, E.S. 1979. N-methyltyramine a boilogically active amine in *Acacia* seeds. Phytochemistry 18:2022 - 2023.

Farnsworth, N.R. 1984. The role of medicinal plants in drug development. In: Krogsgaard-Larsen, P., Brøgger Christensen, S. & Kofod, K. (Eds) Natural Products and Drug Development. Munksgaard, Copenhagen, p 17 - 30.

Finnemore, H. & Gledhill, W.C. 1928. The presence of cyanogenetic glucosides in certain species of *Acacia*. Aust. J. Pharm. 9:176 - 178.

Freudenberger, M.S. 1988. Contradictions of gum arabic afforrestation projects. Bull. IGSM 16:87 - 122.

Führer, H. 1971. Mimosa. Les corps de mimosa et leur utili dans la parfumerie moderne. Dragoco Report 11:219 - 223.

Glicksman, M. 1983. Gum Arabic (Gum Acacia). In: Glicksman, M. (Ed) Food Hydrocolloids, vol. II. CRC Press, Inc., Boca Raton, p. 7 - 29.

Hagos, M., Samuelsson, G., Kenne, L. & Modawi, B.M. 1987. Isolation of Smooth Muscle Relaxing 1,3 diaryl-propan-2-ol derivativs from Acacia tortilis. Planta Med. 53:27 - 31.

Hagos, M. & Samuelsson, G. 1988. Quantitative determination of quracol A,B and (+)-fisitinidol in bark and gum of Acacia tortilis. Acta Pharm. Suec. 25:321 -324.

Hammond, A.C. et al. 1989. Prevention of leucaena toxicosis of cattle in Florida by ruminal inoculation with 3-hydroxy-4(1H)-pyridone-degrading bacteria. Am. J. Vet. Res. 50:2176 - 2180.

Hartley, W.J. 1978. Chronic Phalaris Poisoning or Phalaris Staggers. In: Keeler, R.F., van Kampen, K.R. & James, L.F. (Eds) Effects of Poisonous Plants on Livestock. Academic Press, N.Y., p 391 - 393.

Heyneman, D. 1984. Foreword. In: Lemma, A., Heyneman, D. & Silangwa, S.M. (Eds) Phytolacca dodecandra (Endod). Tycooly Int. Publ. Ltd., Dublin, p ix - xiii.

Hill, D.C. 1973. Chronic cyanide toxicity in domestic animals. In: Nestel, B. & MacIntyre, R. (Eds) Chronic Cassava Toxicity. International Development Research Centre, Ottawa (IDRC - 010e), p 105 - 111.

Huisman, J., van der Poel, T.B.F. & Liener, I.E. (Eds) 1989. Recent advances of research in antinutritional factors in legume seeds. Pudoc, Wageningen.

Imbesi, A. 1964. Index Plantarum, quae in omnium populorum pharmacopoeis sunt adhuc receptae. Messina.

Indikumana, D. & Danilkovich, A.G. 1988. Extracts from a tannin-containing raw material from the Republic of Burundi. Kozh.-Obuvn. Prom - st. 10:73 - 75.

Janzen, D.H. 1974. Swollen-Thorn Acacias of Central America. Smithsonian Contributions to Botany No. 13:1 - 131.

Joseleau, J.P. & Ullmann, G. 1985. A relation between starch metabolism and the synthesis of gum arabic. Bull. IGSM 13:46 - 54,

Jourbert, F. 1983. Purification and Properties of Proteinase Inhibitors from *Acacia sieberana* (Paperbark Acacia) seed. Phytochemistry 22:53 - 57.

Khalil, S.K.W. & Elkheir, Y.M. 1975. Dimethyltryptamine from leaves of certain *Acacia* species of Northern Sudan. Lloydia 38:176 - 177.

Kingsbury, J.M. 1964. Poisonous plants of the United States and Canada. Prentice-Hall, Englewood Cliff.

Kortt, A.A. & Jermyn, M.A. 1981. Acacia Protein Inhibitors. Purification and Properties of the Trypsin Inhibitors fron *Acacia elata* Seed. Eur. J. Biochem. 115:551 - 557.

Langkamp, P. 1987. Germination of Australian Native Plant Seed. Amira/Inkata Press, Melbourne.

Lean, I.J., Anderson, M., Kerfoot, M.G. & Marten, G.C. 1989. Tryptamine Alkaloid Toxicosis in Feedlot Sheep. J. Am. Vet. Med. Assoc. 195: 768 - 771.

Lock, J.M. 1989. Legumes of Africa - a Check-List. Royal Botanic Gardens, Kew.

Luckner, M. 1984. Secondary Metabolism in Microorganisms, Plants, and Animals. 2nd Rev. Ed..Springer-Verlag, Berlin.

Lundgren, B. & Nair, P.V.R. 1985. Agroforestry for soil conservation. In: El-Swaity, S.A., Moldenhauer, W.C. & Lo, A. (eds) Soil Erosion and Conservation. Soil Conservation Society of America, Ankeny, p 703 - 717.

Majak, W. et al. 1980. Seasonal variation in the cyanide potential of arrowgrass (*Triglochin maritima*). Can. J. Plant. Sci. 60:1235 - 1241.

Majak, W. & Cheng, K.J. 1984. Cyanogenesis in bovine rumen fluid and pure cultures of rumen bacteria. J. Anim. Sci. 59:784 - 790.

Maslin, B.R., Conn, E.E. & Dunn, J.E. 1987. Cyanogenic Australian species of *Acacia*: A preliminary account of their toxicity potential. In: Turnbull, J.W. (Ed) Australian Acacias in Developing Countries, Australian centre for Internatioanl Agricultural Research, Canberra, p 107 - 111.

Maslin, B.R., Dunn, J.E. & Conn, E.E. 1988. Cyanogenesis in Australian species of *Acacia*. Phytochemistry 27:421 - 427.

Misra, B.K., Singh, S.P. & Barat, G.K. 1981. Ox-Dapro: the *Lathyrus sativas* neurotoxin. Qual. Plant. - Plant. Food Hum. Nutr. 30: 259 -270.

Mott, K.E. (Ed) 1987. Plant Moluscicides. John Wiley & Sons, Chichester.

Nartey, F. 1978. Manihot Esculenta (Cassava), Cyanogenesis, Ultrastructure and Seed Germination. Munksgaard, Copenhagen.

Naves, Y-R. 1974. Technologie et Chimie des Parfumes Naturels, Essences Concretes, Resinoides, Huiles et Pommades aux Fleurs. Masson et Cie - Editeurs, Paris.

New, T.R. 1984. A Biology of Acacias. Oxford University Press/La Trobe University Press, Auckland/Oxford.

Ngarmsak, S. 1977/78. Cassava-Rice straw complete rations with supplemented urea for dairy cattle. II. In vitro detoxification of cyanogenic glycosides (rate of detoxification of HCN by rumen microorganisms in vitro). Royal Vet. Agric. Univ., Veterinary Faculty for FAO Fellows, Copenhagen.

Nicholson, S.S. et al. 1989. Delayed Phalaris Grass Toxicosis in Sheep and Cattle. J. Am. Vet. Med. Assoc. 195: 345 - 346.

Nongonierma, A. 1977. Contribution á l'étude biosystematique de genre *Acacia* Miller en Afrique occidentale. Bull. I.F.A.N., ser. A 39:23 - 74 (Edaphologie) and 318 - 339 (Distribution bioclimatique des différents taxa).

Ory, R.L. (Ed) 1981. Antinutrients and Natural Taxicants in Foods. Food and Nutrition Press, Westport.

Pettigrew, C.J. & Watson, L. 1975. On the Classification of Australian Acacias. Austr. J. Bot. 23:833 - 847.

Peyron, L. 1972. Acacias in Perfumery. Mimosa and Cassie ancienne. Am. Cosmetics and Perfumery 87:37 - 41.

Price, D.A. & Hardy, D.V.M. 1953. Guajillo Poisoning of Sheep. J. Am. Vet. Med. Assoc. March:223 - 225

Raymond, W.D. 1951. The use of Acacia pods and bark as tanning materials. Colonial Plant and Animal Products 2(4):285 - 291.

Rexen, F. & Munck, L. 1984. Cereal Crops for Industrial Use in Europe. The Carlsberg Research Center for The Commision of the European Communities (EUR 9617. EN), Directorate - General Information Market and Innovation, Luxembourg.

Rimington, C. 1935. The occurrence of cyanogenetic glucosides in South African species of Acacia. Onderstepoort J. Vet. Sci. Anim. Ind. 5:445 - 464.

Rodman, E.R. 1978. Glucosinolates, methods of analysis and some chemosystematic problems. Phytochemical Bulletin (Phytochem. Sec., Botanical Soc. of Am., Inc.) 11, 6 - 32.

Rosling, H. 1986. Cassava, Cyanide, and Epidemic Spastic Paraparesis - a study in Mozambique on dietary cyanide exposure. Dissertation Uppsala University, Sweden. Acta Universitatis Upsalensis no 19.

Rosling, H. 1988. Cassava Toxicity and Food Security - a review of health effects of cyanide exposure from cassava and of ways to prevent these effects. Unicef (African Household Food Security Programme)/International Child Health Unit, University Hospital, Uppsala.

Seif, El DIN AG 1975. The future of gum arabic in Sudan. Sudan International 1 (12-13):24 - 26.

Seigler, D.S. & Conn, E.E. 1982. Cyanogenesis and Systematics of the Genus *Acacia*. Bull. IGSM 10:32 - 43.

Seigler, D.S. & Ebinger JE 1987. Cyanogenic glycosides in ant acacias of Mexico and Central America. Southwest. Nat. 32:499 - 504.

Seigler, D.S. & Ebinger, J.E. 1988. Acacia macracantha, Acacia pennatula, and Acacia cochliacantha (Fabaceae:Mimosoideae) species complexes in Mexico. Syst. Bot. 13:7 - 15.

Shakla, N.K., Verma, P.C. & Aswal, S.S. 1987. Physical and Mechanical Properties of Acacia melanoxylon and Castanopsis indica. J. Timber Dev. Assoc. India 33:9 - 17.

Sheppard, J.S. & Bulloch, B.T. 1986. Management and Uses of *Acacia* spp. (Wattles) and *Albizia* Spp. (Brush Wattles). In: van Kraayenoord, C.W.S. & Hathaway, R.L. (eds) Plant Materials Handbook for Soil Conservation. Vol. 2 - Introduced Plants. Water and Soil Miscellaneous Publication No. 94. Water and Soil Directorate, Ministry of Works and Development, Wellington, p 7 - 19.

Skarpe, C. & Bergström, R. 1986. Nutrient content and digestibility of forage plants in relation to plant phenology and rainfall in the Kalahari, Botswana.J. Arid Environments 11:147 - 164.

Skerman, P.J. 1977. Leguminous Browse. In: Skerman, P.J. Tropical Forage Legumes. FAO, Rome, p 431 - 525.

Skolmen, R.G. 1986. Acacia (*Acacia koa* Gray). In: Bajaj YPS (Eds) Biotechnology in Agriculture and Forestry 1. Trees I. Springer-Verlag, Berlin, p 375 - 384.

Summerfield, R.J. & Roberts, E.H. (Eds) 1985. Grain Legume Crops. Collins, London.

Tameim, O., Zakara, Z.B., Hussein, H., Gaddal, A.A. & Jobin, W.R. 1985. Control of Schistosomiasis in the New Rahad irrigation Scheme of Central Sudan. J. Trop. Med. Hyg. 88:115 - 124.

Turnbull, J.W. (Ed) 1986. Multipurpose Australian trees and shrubs: lesser-known species for fuelwood and agroforestry. Australian Centre for International Agricultural Research, Canberra.

Turnbull, J.W. (Ed) 1987. Australian Acacias in Developing Countries. ACIAR Proceedings no. 16. Australian Centre for International Agricultural Rasearch, Canberra.

Vassal, J. 1972. Apport des recherches ontogéniques et séminologiques a l'étude morphologique, taxonomique et phylogénique de genre Acacia. Bull. Soc. d'Histoire Naturelle de Toulouse 108:125 - 247.

Vassal, J. 1983. Gommiers et production gommiere. In: Acquisitions recentes dans les domaines des hydrocolloides vegetaux naturels. Presses Universitaires d'Aix-Marseille, p 5 - 17.

Vassal, J. 1983. Interet des Acacias pour l'amenagement de la Corse. Etude generale et experimentation. Publication SOMIVAC.

Vassal, J. 1985. Gummiferous acacias and gum productivity, some aspects of current research. Bull. IGSM 13:30 - 37.

Way, J.L. et al. 1988. The Mechanism of Cyanide Intoxication and its Antagonism. In: Cyanide Compouns in Biology. John Wiley & Sons, Chichester, p 232 - 243.

92

Webbe, G. 1987. Molluscicides in the control of schistosomiasis. In: Mott, K.E. (Ed) Plant Molluscicides. John Wiley & Sons LTD, Chichester, p 1 - 26.

Weder, J.K.P. 1981. Protease Inhibitors in the Leguminosae. In: Polhill, R.M. & Raven, P.H. (Eds) Advances in Legume Systematics, part 2. Royal Botanic Gardens, Kew, p 533 - 560.

Weder, J.K.P. & Murray, R. 1981. Distribution of Proteinase Inhibitors in Seeds of Australian Acacias. Z. Planzenphysiol. 103:317 - 322.

Worthing, C.R. & Walker, S.B. (Eds) 1987. The Pesticide Manual, A World Compendium, 8'th Ed. British Crop Protection, Thornton Heath, p. 1 - 851.

Yamamoto, H-A. 1989. Hyperammonemia, increased brain neutal and aromatic amino acid levels, and encephalopathy induced by cyanide in mice. Toxicol. Appl. Pharmacol. 99:415 - 420.

The Mali Project at the University of Oslo

Tor A. Benjaminsen

Background

The Mali-project forms part of the SSE-Programme, a development and research programme for countries in the SSE-belt (Sahel, Sudan, Ethiopia), financed by the Norwegian Ministry of Development Cooperation.

This research project is based on an integrated interdisciplinary collaboration with common goals. For practical purposes the research activities are organized within three sub-projects, each being staffed by Malian and Norwegian researchers from various disciplines. These sub-projects are:

1. Pastoralism and the utilization of resources
2. The use of wild plants for food, medicine and handicrafts.
3. Food security and nutrition at the household-level, and the role of women in the management of natural resources and food security.

At the University of Oslo, researchers come from the following departments: the Department of Geography, the Council for Environmental Studies, and the Institute of Nutrition Research. The Malian researchers are from several national schools and research institutions representing both the natural and social sciences.

Important goals for the project are:

1. Upgrading of research expertise in Mali and Norway in areas linked to the research themes.
2. Applied research as a tool for better planning.
3. Dissemination of results from research to both local populations in the field and to a broader public in Mali and in Norway.

The following is a presentation of the sub-project "Pastoralism and the utilization of resources". The Norwegian participation is from the Department of Geography at the University of Oslo; represented by professor Just Gjessing, graduate student André Kammerud and myself. The research on water resources of professor Lars Gottschalk at the Department of Geophysics will also be integrated in the project. On the

Malian side two geographers, two ecologists, one hydrologist and one sociologist are participating. In addition, several Malian graduate students are expected to do their theses as a part of this project.

Gourma: The field area

The Gourma region was chosen as the geographical setting for the research activities. This is also an area in which the Norwegian Church Aid is engaged in a major development project.

Gourma is a natural region bounded by the interior delta of the river Niger to the West, the bend of the river to the North and East and the republics of Burkina Faso and Niger to the South. The region was seriously affected by the Sahelian droughts of the 1970's and 1980's.

Pastoralism in different forms, from the pure nomadic to agro-pastoral versions, is the main activity in this region. The resource base is not suited to many other activities apart from herding, with the exception of the Niger valley where rice is cultivated and the southern parts of Gourma where millet is grown. Livestock is important everywhere.

The population of Gourma consists of several ethnic groups like Tamasheqs (Tuaregs), Maures, Songhays and Fulanis.

An Integrated and Interdisciplinary Approach

In Gourma, environmental and social and economic development are closely linked. The study of the interrelationships between different aspects of the social, economic and ecological changes demands an integrated approach. The processes related to natural and social sciences are interlinked in complex cause-and-effect-systems. It is therefore necessary to study the interrelationships between environment and development in an integrated and interdisciplinary perspective. This is accomplished by close collaboration between researchers posing common goals.

The work in the sub-project concerned with geography and ecology is divided into four studies so far:
1. Resource monitoring
2. Demography
3. Land use
4. Wood fuel consumption for cooking

1. Resource monitoring

Since 1984, the International Livestock Center for Africa (ILCA) has been monitoring changes in the herbaceous and woody vegetation in Gourma. On selected sites, the evolution of the vegetation is monitored through the seasons and from year to year. Through this project, which is under the leadership of the ecologist Dr. Pierre Hiernaux, a series of interesting data on ecological processes taking place in the Sahel has been collected. If one is to do research on changes in the resource base in Gourma, it is important to build on ILCA's experience in the area. Cooperation with ILCA therefore forms an important part of the project. This cooperation consists of an expansion of monitoring of natural resources to include other resources such as water and livestock which are important for pastoral production in Gourma.

Together with ILCA and the local administration represented by "le Service de l'Elevage", which is responsible for the management of the pastoral resources, we have established a network of research sites. The work is concentrated in "le Cercle de Gourma Rharous" which is an administrative unit covering about half of Gourma. In each of the seven "Arrondissements" in the "Cercle", three sites are established where local agents of the "Service de l'Elevage" are responsible for data collection.

The fact that the local administration takes part in this collaboration is an advantage in many aspects. It is this administration that has the responsibility for resource management in the area, it is also this administration that is likely to make use of the information generated by this research.

It is therefore important that the local administration is actively integrated in the research process. In this way the local resource managers can acquire new experience and develop competence within the area. The project will also profit from having fieldworkers who represent a certain continuity and who have an impressive history of field experience.

Measurements on the 21 sites will be taken three times a year; at the end of the rainy season (September/October), after the dry and cool season (February) and at the end of the dry and hot season (May). Vegetation from the selected sites is harvested and weighed to quantify the changes occurring during the annual cycle, in addition to those taking place from year to year. Changes concerning species will also be studied. By choosing sites with varying pressure from humans and animals, it is possible to study the impact of different intensities and types of grazing on the grass cover.

96

The methodology concerning the monitoring of water and animal resources is also being elaborated upon. One hydrologist and one zoo-ecologist from Mali are responsible for this work.

A common methodology for monitoring the pasture resources of the Sahel is the application of NOAA-satellite images. This will also be used in our project. M.A.-student André Kammerud is studying the impact of different soil types in Gourma on biomass estimations using NOAA-images. From this study, we hope to make corrected biomass maps showing the biomass at its maximum at the end of each rainy season.

2. Demography

For several reasons it is not easy to obtain reliable demographic data from Gourma. First of all the population is nomadic. There is, in other words, a great mobility in the population resulting from changes in the resource base. Another problem is taxation in Mali is per capita. In consequence, suspiciously few children are declared during the national census. In a demographic study, one therefore has to cope with a certain suspicion and reservation on the part of informants (even though names of individuals are not noted). Other problems are related to age identification, the gathering of data on child mortality or on spacing between children in a family. One has to find methods by which these problems may be solved, in order to get reliable demographic data.

An experienced Malian geographer is responsible for the demographic study. His first step is to register all the existing data in the local tax registers. Afterwards, he will check these data against case-studies, to get the best picture possible of the demographic situation in Gourma.

In Gourma a significant level of emigration, and increased mortality rates are evident for the periods of drought in the 1970's and 1980's. According to official records, the population of Gourma decreased by about 1%, from 1976 to 1987. It would be interesting to know what is happening now, after two fairly good rainy seasons; for instance whether or not people are moving back to their former areas of residence.

Local authorities and NGO's in Gourma are very interested in getting reliable demographic data from the area, which are important as a basis for future planning. It would be interesting to relate the demographic data to that from the resource monitoring; especially to get an overview of supply and demand ratios.

3. Land use

Gourma can be divided into four zones according to production system: The river valley (cultivation of rice, agro-pastoralism), "the lakes"; the northeastern part of the Niger inland-delta with mostly dry lakes (millet cultivation, agro-pastoralism), the interior of Gourma (nomadic pastoralism) and lower Gourma (millet cultivation, agro-pastoralism).

These four zones cannot be studied separately, because there is considerable complementarity between them arising from the fact that the different zones are used by the people during different periods of the year. To begin with, the project will nevertheless concentrate on land use in the river valley and in the interior of Gourma.

There is also an important interdependence between these two zones. A large part of the nomadic population has always been dependent on water and pastural resources in the valley during the dry season. Several nomadic groups have also received confirmation, from French and later Malian authorities, of their rights to use the land in the valley. Nevertheless, there are numerous conflicts concerning rights to use the land between nomads (Tamasheqs) and sedentary Songhays. In fact, maybe as much as half of all land in the valley is left unused because of these conflicts.

One of our tasks is to map land use in the Niger valley. In addition, we will identify which groups possess the right to use the land and which groups claim this right. We will also study the distribution of the annual crop production, which groups represent the working force and who takes the decisions. It may also be interesting to observe how much rice is consumed by the unit of production itself and how much is sold to local merchants (to be bought back later at many times the original selling price).

We also intend to map which nomadic groups use the different pastures and water resources in the interior of Gourma throughout the year. In this connection, the changes in land use over time, from the colonization of Gourma to the present day should be studied. In this way, one could show some of the background for the lack of management evident today and provide some inspiration for a new system of sustainable resource management.

In this connection, it may be mentioned that in pre-colonial times, the Imoushars (Tamasheq warriors) governed most of Gourma and controlled the use of natural resources. Management was based on an authoritarian system in which the strongest group controlled access to the most attractive resources. Imoushar control over Gourma was however gradually reduced during more than 20 years of war with the French, and in 1916 they suffered a fatal defeat. The new power structure eventually led to a less controllable utilization of the resource base, and consequently more open land use rights gradually arose, and large areas became available for common use.

A closer look at what has actually happened to land use rights and land use in Gourma since colonization would be useful for developing a new system for the sustainable use of resources. In this connection it would also be very interesting to see which parts of the old system survive, as a new system could build upon the remains of the traditional management principles and structures.

4. Wood fuel consumption for cooking

This is a study concerning the consumption of wood fuel for cooking in various villages in the river valley. The valley has been selected because it is here that we find the largest sedentary population in Gourma. In the nomadic areas one cannot talk about energy problems, since there is enough dead wood on the ground. Around the villages, one will nevertheless have to go farther and farther to collect dry wood. There are women who walk up to 10 km each way two or three times a week to collect wood. There is still very little tree cutting for woodfuel.

The consumption of wood fuel, charcoal and other sources of energy (e.g. cow manure) in several villages will be studied. Other aspects under study will be the number of wood-saving stoves, how much of the wood fuel consumed is collected by the consumers and how much of it is bought, the amount of time spent in collecting it, who collects it for their own consumption and who are professional collectors, the areas in which wood is collected, which species are preferred as wood fuel and which are less preferred and what measures the local population proposes to meet an energy crisis.

Two questionnaires have been worked out. One is intenden for the women who are responsible for cooking in the individual households. The other is intended for those who sell wood fuel and charcoal. Interviews will take place in the Songhay and Tamasheq languages, and will be carried out primarily by female interviewers.

Research for Planning

As mentioned earlier, an important goal for the project is that the research results can be used by the local authorities or NGO's in planning for a better future in Gourma. An overall goal for our sub-project is to give a proposal for a land use plan for central parts of the field area.

This can only be accomplished by viewing the data in a holistic context and by finding methods by which to integrate the different data. To understand and present relationships between different factors like vegetation and water resources, herd composition and number of animals, demographic data and other social and economic variables in relation to production and land use, it may be useful to apply software such as Geographical Information Systems (GIS).

Agro-Pastoral Society

Ebbe Poulsen

When speaking of development projects it is normal practice to distinguish between different types of projects and, in many cases as a consequence, also between different types of target groups. Presumably these classifications are made for technical reasons - a fact which is largely reflected in the way many international development agencies are structured with e.g. separate divisions for fisheries, livestock production, agriculture etc. But many of these divisions also have a background, knowingly or not, in classifications established within evolutionary inspired historical theory, dating back to the end of the last century.

A classification widely used all over Africa, as well as elsewhere, in connection with development projects distinguishes between pastoralists, semi-pastoralists and sedentary agriculturalists. Were this distinction purely academic, it would only be of little interest here, but unfortunately it has had a strong impact and has given rise to a large number of development projects which use this distinction when stating their objectives, thus establishing specific development projects for pastoralists, agriculturalists etc.

Unfortunately, reality is a little more complex than this and in actual life there is a close interrelation between the various groups of pastoralists, agriculturalists etc. The connection plays a significant role in the way development is perceived as well as in the outcome of a specific development process.

In the following, I shall try to demonstrate this complexity in general terms although also referring to specific Somali and East African contexts.

The starting-point is the assumption that no nomadic or pastoral society has in historic times been self-sufficient in the true sense of the word, but has always relied heavily on agricultural societies. Of course, this dependence can be expressed in purely economic terms. We know, for instance, that no pastoral society can subsist directly and exclusively on its pastoral production but is forced to engage in exchange with its agricultural neighbors. Although there does seems to have been a tendency to play down this exchange in most monographs concerning pastoral societies in order to make them resemble closed, self-sustaining societies fit for functional or structural analysis, the exchange between pastoral and non-pastoral societies is nevertheless a historical fact which must be taken into consideration in order to explain the potentials for social change inherent in pastoral societies.[1]

The forms of economic dependence have of course varied greatly from society to society and have changed radically through history. Usually the volume of the goods exchanged has been rather small - what counts is the strategic nature of the goods involved which again implies that the social aspects of the economy must be stressed. Concerning the Somalis, it seems that before World War II, necessities such as clothes, sugar and tea were essentials which had to be obtained from other societies, but an undetermined amount of grain and other staples had to be brought in from outside as well.

The inter-dependence of pastoral and non-pastoral societies has thus not been purely economic but has also implied several social and cultural bonds. Many of these were necessary just in order to engage in the economic exchange in itself and they have taken many different forms through history. Basically it seems, though, that in order to establish and maintain trading relations some sort of political or military superiority had to be invoked on the part of the pastoralists. In the few cases where the pastoralists have not been able to exercise this superiority, it seems that they have had to submit altogether to the sovereign power of a strong and centralized political leadership on the part of the non-pastoral society.[2] In all cases, there seems to have been a clear hierarchical relationship between pastoralists and non-pastoralists.

---

1. The classical example of this is of course E. Evans-Pritchard's works on the Nuer. But even the young Evans-Pritchard had to admit that the cattle-herding Nuer could not subsist without either engaging in agricultural production themselves or obtaining agricultural produce from their neighbors through barter.
2. An example of this could be seen in North-western Tanzania, where the pastoral population to a very high degree was subject to the various kings and chiefs in the area who even owned and controlled vast parts of their herds. (See Poulsen, E.: 1980)

Surplus Population

Another noteworthy fact is that most pastoral societies have constantly produced a surplus of persons who have been pushed off to agricultural or urban centers. This push-off has in general, consisted of two distinct groups, i.e. the very rich, typically an elderly rich man who has settled together with most of his close family, living off former investments or maybe engaging in commercial activities, and the very poor who most often have been pushed off as individuals and struggle to survive as casual laborers in distant towns or agricultural areas. To explain the social process which leads to this constant 'push-off' from the pastoral societies, it is necessary shortly to explain some basics of the social structure of most of these societies.

Basically, most pastoral societies are constituted around a quite small nucleus - very often only an extended family - which forms the main productive unit. Relationships within this nucleus are often characterized by a rather sharp division of labor based upon sex and age. Relations between nuclei are generally expressed in terms of kinship, but in all pastoral societies, there seem to be certain non-kinship based associations as well, invoking relationships through friendship, age etc. Given a background where production is a high-risk enterprise due to a harsh environment and excessive competition between productive units, the various social relationships to a certain degree work as a kind of social assurance, securing the individual in the productive system. But this is true only to a certain degree. The same relationships, based on more or less well-defined rights and obligations, are also vital instruments in the pushing-off of members. Typically this can happen one of the following ways:

A family or group of families loses - due to bad grazing, illness among the animals or productive members of the family, theft or other common hazards - the main part of their herd. In order to survive, they borrow a number of new animals from kin or through other relationships. Normally, such a loan will enable the family to rebuild their herd within a few years and they will be able to pay back their debt. But if they suffer another stroke of bad luck - especially within those few first years - they may not be able to borrow animals again or, which seems to be more common, they can only borrow on quite severe terms. This means that they might only be able to keep a minor part of the offspring of the borrowed animals; that they cannot compose their flock according to their own needs and abilities and that they might therefore slowly be excluded from the social network among the other pastoralists. The persons involved in this will probably themselves be able to survive within the pastoral society, but for the next generation, there will be problems. If a son wants to marry, he needs animals for the brideprice as well as for the mandatory gift to the bride. Even if he is lucky enough to have a sister who has recently married, he will probably not be able to collect a sufficient amount of animals for the wedding, as most of his sister's brideprice will already have been claimed by the original loaners of the animals. So his only chance to obtain a brideprice will either be by earning it himself - which nowadays usually means taking wage labor outside the pastoral community - or by concluding a contract with a more fortunate relative or acquaintance obliging him to work as a herdsman in return for some vague promise of a solution to his brideprice problem.

Other families might not be as fortunate as the one just described. Each year a number of families are broken up altogether due to loss of animals or productive family members. Here, the male individuals will often either be pushed-off from the pastoral production or subsist as hired herdsmen. Unmarried girls will usually be taken care of by relatives in whose households they will work as unpaid domestic servants until marriage, when the adoptant usually will withhold at least a part of the brideprice. Married women on their own have a fair chance to re-marry but often, they too will try to leave the pastoral area.

This is the way most of the push-off from what could be called the lower end of society takes place. From the top, an interrelated process takes place.

The Somalis say that a rich man is a man who is able to give. Giving in this sense is to a very large extent a replacement for the term to lend. As we have seen, more fortunate families which have a surplus to lend to relatives and others in need can to a large extent profit from this loan.[3] The risk that of loss is high - probably much higher than most modern banks would accept - but some lenders are fortunate enough to profit heavily from the loan. This means first, that their herds grow far beyond their own subsistence needs and second, that most of the herding involved can now be done by persons from outside the original nucleus of production. To obtain the maximum profit from the herd - and at the same time to be able to enjoy this profit -

---

3. This is not to say that most lending of stock is done an eye to potential profit. On the contrary, I believe that most of the lending described above is based on much more altruistic motives; but it remains a fact that the pay-off can be quite substantial.

the family will now usually turn its attention towards other aspects of production, e.g. the buying and selling of produce. To do this one must leave the pastoral area and move to a location in - or between - the main market centers, that is to say the major towns.

Those pushed off from the pastoral society thus form two distinct groups. And although they have very little in common both of them, though physically absent from the pastoral areas, remain part of pastoral society: they exert influence there, and influence is exercised over them. But of course our elderly well-off gentleman is by far the more influential, so we will turn our attention to him first.

Typically, he will retain great economic interests in the pastoral areas. He may be the owner of large herds, attended to by either hired laborers or kin, paid in kind according to a fixed contract. As he has no immediate subsistence needs to be covered in kind by the flock, he will seek to change flock composition to maintain the optimal marketable composition in relation to labor availability and costs. In Somalia and Kenya, this means meat production, normally in the form cattle, less often camels.

Flocks will thus tend to be composed of either camels or cattle. The special mixture of cattle, camels and smaller animals - which, as we shall see later, is predominant in the "purer" form of subsistence production - has been given up; at the same time there will be a tendency to keep the flocks rather young and with equal representation of both sexes. The aim being to produce the maximum volume of meat in a minimum of time and - although not as crucial - with a minimum of labor in-put there is a tendency to increase the actual number of animals even at the cost of a decrease in the quality and size of individual animals. This tendency is sustained by the wide-spread practice of pricing animals by their number and not by their size and quality.

Ecological Consequences

Ecologically, this type of production places rather one-sided demands on the surroundings. Due to the uniformity in flock-composition, the exploitation of the surroundings will be rather monotonous and due to the number of animals and the proportionally large share of animals not yet mature, food intake will be very high. As the surroundings can only partly be exploited, this kind of nomadism requires proportionally more land per energy unit produced. Although I am no specialist, it appears to me, that the transformation of the vegetation due to this kind of production is severe because the homogenous flocks choose rather few species for subsistence, leaving the rest to their "natural" growth. Large parts of the vegetation thus remain uncontrolled, which in certain instances may cause undesirable regrowth of species which will eventually overtake the exploitable ones, thus destroying the feeding potential of the area. Where flocks are composed exclusively of cows, the result may be an undesired growth of scrub and thorny trees, while camel flocks will usually subsist by browsing, thus opening the foliage cover and offering grasses and low scrub better conditions for growth. In both cases, the result might be destruction of the productive capacity of the area for that specific animal.

On the other hand, due to the size of the flocks, a more common danger may be total devastation of the productive capacity when roots, bark etc. are be destroyed in consequence of the mere presence of great numbers of animals. This is seen at a number of places in Somalia and Kenya, mainly near the large towns, and the animals involved are most often cows.

Apart from the ecological and class-related consequences the push-off also implies a strong social interaction and interdependency between various other social groups. First of all, it is obvious that the rich herd-owner must keep in constant contact with the pastoral community in order to continue the exploitation of his less fortunate kinfolk. As mentioned, it is essential that he can continue to fulfill the role of the one who gives and, as society grows more complex, to give is not only a question of giving animals but extends to political influence vis-á-vis the modern state. A modern, multifaceted exploitative patron-client relationship is thus established.

Although our elderly, rich gentleman derives most of his wealth from the pastoral sector it is obvious that his production is dependent on the market and thus open to several kinds of political and economical intervention from the state. In order to steer his way through this, the "absentee herd-owner" will need to create alliances and obtain influence within the leading economic and political groups. The internal linkage between these groups are more often than not based on various business transactions and the "absentee herd-owner" will have to comply with that.

As already noted, herding is a high-risk enterprise; therefore, most rich herd-owners choose to spread their investments over several types of business. Many invest in agricultural land, usually not for immediate profit but as a long-term security. For much of land owned in this manner, the real value is not its productive capacity but its future market value. Large tracts of farming land thus lie idle or underutilized because the owner has neither the considerable capital input necessary to make it an investment competitive with other productive investments nor a genuine interest in commercial farming.

In eastern Africa, the most popular sector of productive investment at the moment seems to be the transport sector. Much of the surplus derived from the pastoral sector is thus bound to be used for the purchase of small pick-ups, busses and trucks instead of being re-invested in the livestock sector. For some, the livestock sector thus becomes a source of wealth, readily exploited without considering the consequences in terms of sustainability.

Not only the rich have a role to play in the interaction of the various economic sectors of society. For numerous poor pastoralists arable agriculture becomes the only realistic alternative to small-scale pastoralism or a life as paid herders. Very often, though, arable agriculture is only one of several strategies employed by one specific family unit as it will try to diversify its economic opportunities. In most cases, therefore, there is a very close relationship between a newly settled agricultural household, various pastoral households and very often a number of urban households as well.

The role of the agricultural household in this web of mutual inter-dependence is ideally to secure as large a part of basic staples as possible for the entire group. This means that the production will be subsistence oriented rather than market-oriented, even in times of surplus. Market oriented mechanisms to increase the agricultural production as recommended by the World Bank (among many others) are therefore likely to have little impact on such forms of production.

The close connection between the agricultural and the pastoral sector also allows investment to flow freely between the two. Usually, this means that any surplus produced in the agricultural sector will be absorbed into the pastoral sector where returns tend to be considerably higher than in arable agriculture. In other instances it means that agriculture is viewed as a convenient place for non-productive members of the pastoral society, i.e. infants, schoolchildren and the elderly.

In both cases, surplus is drained from arable agriculture in order to replace the surplus originating there but diverted into various market oriented activities. And the result, as in the pastoral sector, is lack of investment to secure efficient production on a sustainable basis.

International Provenance Trials

DANIDA Forest Seed Center

Follow-up Teak and Gmelina international provenance trials

Nearly 20 trials of teak and 30 trials of gmelina were included in the assessment and the results were published in "Evaluation of an International Series of Teak Provenance Trials" by Keiding, Wellendorf and Lauridsen, 1986, and "Evaluation of an International Series of Gmelina Provenance Trials" by Lauridsen, Wellendorf and Keiding, 1987.

Preparations for the second phase

Two circular letters have been distributed to the host countries outlining the programme for consolidating evaluation, and asking for collaboration. A limited but representative number of trials will be selected for a second evaluation, depending on the response to the questionnaires. The resource requirement is estimated on the basis of evaluation of 10 trials of each species. A few representative trials will be used for pilot assessments, checking the applicability of the current scoring system. Information will be given to the collaborating institutes outlining the procedures to be applied and agreements established regarding the fieldwork and local collaboration.

Assessment and evaluation

Similar procedures as applied at the first evaluation will be used. The results will be published by DFSC.

Guidelines for the selection of seed sources and seed supply etc.

Recommendations for the choice of seed sources will, if applicable, be issued. A list of available seed sources will, if possible, be published with references to potential seed suppliers. Collaboration with host countries will be established for the development of techniques applicable to the vegetative propagation of superior provenances.

Central American Pines

During 1972-74 the Commonwealth Forestry Institute, Oxford, established an international series of provenance trials with Central American Pines in 40 countries. In many countries the results for certain species and provenances were so promising that a great demand for seed was created. During the project period 1981-86, DFSC undertook the task of collecting and supplying seed of the most promising seed sources for the establishment of seed stands and conservation stands. Seed is also available for research purposes and for the establishment of pilot plantations. DFSC has supplied approx. 5,000 ha stands. DFSC still has about 375 kg of this seed in stock, and it is essential that this valuable seed is used for these purposes.

Seed samples of the most important sources will be kept for medium to long-term storage.

Seed supply

Availability of seed will be announced through seed stock lists, circulated on request internationally. Recipient countries should indicate the purpose of their request (research samples, stand establishment etc). New collections of Central American Pines are not anticipated.

Establishment and development of stands

Advice will be given on the establishment and management of the stands. Recipient countries will be requested to inform DFSC on location, size and development of stands. If feasible, DFSC will endeavour to follow the development of the stands. Opportunities will be taken to visit and appraise stands in cooperation with host countries to facilitate discussion of further development procedures aimed at maximum seed production.

As information is made available from the collaborating countries, a list of stands available for future seed supply will be circulated with approval from the host countries. In such cases seed may become available at cost price from the countries in question.

South East Asian Pines - Pinus kesiya and Pinus merkusii

At a IUFRO conference in 1984, DFSC participated in a decision to undertake an international provenance trial of P. kesiya, coordinated by the Oxford Forestry Institute (OFI). DFSC has supported the collection of seed from natural stands in South East Asian countries and OFI has organized collections of local landraces in 4 countries. The seed is stored at the seed provenances (including 17 in which seed from individual trees has been kept separate) and 7 landraces.

The seed will be used for 4 main purposes:

1) to study variation between provenances (bulk seed)
2) to study variation within provenances (single-tree seed)
3) to establish breeding seed orchards (single-tree seed)
4) to establish provenance conservation stands (bulk seed).

Establishment of provenance and single-tree trials

The OFI is in consultation with DFSC coordinating the trials. Requests for seed will reach DFSC through Oxford and be distributed from DFSC directly to the collaborating institutes. Most of the seed will be distributed in 1989-90.

In agreement with the host countries and the OFI, DFSC will take the opportunity, during consultations related to other work, to provide advice on the management of the trials in DANIDA supported countries. Seed not required for this purpose will be stored at DFSC's seed bank.

Seed collection of Pinus merkusii

Supplementary collections of Pinus merkusii for the regional SE Asia programme is expected to be organized by Thailand and supported by DFSC.

Establishment and development of stands for seed supply

The procedure will follow the lines and principles indicated above for Central American Pines.

Seed Biology and Technology

Activities will focus primarily on multipurpose dry zone species which have been, or will be, identified as future priority species. Important indigenous species will be included where the seed problems are clearly a hindrance for propagation.

Specifically, the work will be concerned with:

- establishing routine handling, storage, testing, pretreatment and germination methods for species new to the programme
- developing methods and equipment for drying seed that are less

damaging than those that use heat, i.e. cool drying methods that do not encourage insect infestation

- searching for and adapting simplified methods for the rapid determination of seed moisture content

- continuing the development and testing of storage methods that control insect pests and thus limit the use of (often noxious) insecticides. This development would be particularly relevant in relation to the seeds of African leguminous species which are particularly susceptible to insect damage. Work on this problem might benefit from collaboration with the seed centers in Burkina Faso and Kenya and the Pest Control Institute in Denmark.

- developing long term seed storage methods suitable for application in the tropics

- furthering development of methodology and equipment for the scarification of hard-coated seed

- the continuation of developing methods for handling, transportation and short term storage of seeds characterized by high moisture content, e.g. that of Azadirachta indica, Melia azadirach, Balanites, Zizyphus.

Seed Bank

The seed bank provides seed handling and storage facilities for the above research and development activities, and for field projects supported by DANIDA.

Collaboration with Danish Research Institutes

a) Tree Breeding Institute, the Arboretum Hørsholm
b) Tree Improvement Station, Humlebæk
c) Seed Pathology Institute for Dev. Countries
d) Danish Pest Control Station
e) Botanical Institutes of the Universities of Copenhagen and Århus

Further details will follow when agreements have been finalized.

Research and Development Assistance

Sofus Christiansen

The usefulness of research is generally recognized when development assistance is discussed. Questions mainly relate to the problems of how much research is needed and of what type. To provide a basis for answering these questions, a small survey of recent research paid for by the Danish Development Assistance Agency, DANIDA, is given below, followed by an attempt to indicate future research needs.

Present research related to Danish development assistance

Obviously, most of the scientific findings used in development stem from research carried out in other contexts. Here only the relatively small contributions directly paid for and probably initiated by DANIDA will be mentioned.

A very accurate analysis will require a more substantial insight into DANIDA accounting as well as application of a more rigid definition of research than the present writer has employed. However, the following extracts taken from DANIDA's annual report 1988, supplemented with a few estimates, give a rough estimate of the amounts spent on development aid.

Direct contributions to research (in million DKK):

Research at international institutions (e.g. CGIAR)	20+
Assistance to developing countries' own research	10
Danish research in developing countries	
Center for Development research	10
'free' research	10
Danish research activities in Denmark (Forest	
Seed Centre, Seed Pathology, Bilharziosis)	10

Indirect contributions to research (estimated figures):

Research/study components in projects	10*
Research by World Bank, European Community, etc.	10*
Total (million DKK)	*80*

The effects of estimating and rounding the above figures have possibly resulted in a reduced sum total, since DANIDA reports a total cost of research amounting to 1,5% of the entire amount (6 billion DKK) set aside for development assistance - which would make 90 million DKK. Also, problems of defining activities have probably influenced these rough estimates. Are 'studies', 'trials', 'adaptive development of methods' etc. research or not?

Even when such potential omissions are taken into account, the total funding for research is modest. It corresponds approximately to the a-mount supplied for Danish research by the private Carlsberg Foundation, but represents a much lower percentage of total spending than that invested by innovative firms in research (often 4-5% or more). In this connection it should be noted that the amount reserved for 'free research' (available for the support of individual projects, the themes of which are selected by the researchers themselves) is only 10 million DKK.

Considering the importance of independent, innovative research and possibilities for the recruitment of young scientists the amount seems even more modest. Recruiting and training of young researchers for work in developing countries is a considerable problem, as to provide them with a knowledge of the language, culture etc. specific to the working area chosen often necessitates further funding, i.e. extra expenses for the agencies employing them.

To solve this problem, several approaches may be used. One is to give students in their senior year an opportunity to stay in project areas, attached to a senior researcher/manager as a kind of apprentice, another would be to involve universities/higher educational institutions in research projects in which students can participate.

A preliminary survey of research needs may possibly shed some light on the question of scarcity of means reserved for research in developing countries.

The Demand for Research Related to Development Assistance

Demands are expressed very differently according to type of applicant.

Developing countries: ('recipients') express their needs roughly as follows, (when they define them at all):

- *no research* is required from the outside because:

a) we are capable of doing it ourselves (esp. in socio-economics).

b) we are in more urgent need of the money itself (ie. research tends only to produce paper work and not visible results).

- at least only *specialized research* is needed:

a) to develop our own capacity.

b) to solve specific problems identified by ourselves.

In every case, *efficiency* is required.

In *developed countries* ('donors') several more groups can be identified:

- *Politicians* generally hesitate to express needs for research, since this may result in heavy expenses for activities not necessarily defined by themselves. There is a political need for research, but mainly related to identification of models for development etc.- that is, helping to define policies. But this type of research may be dangerous to support, since it is often critical.
- The *researchers* want plenty of opportunities for *free* research, opportunities which are always in short supply. Resistance to the expansion of funding is often based on two things:

a) the quality of applications/research is often low and

b) its usefulness is not always immediate.

Reasons for this are many. Young researchers often have a rather limited prior knowledge of the situation in "their" developing country and their projects may therefore often be of little interest if they either duplicate previous research, or are too ambitious to be carried to fruition.

Initiatives to improve recruitment of young researchers are obviously necessary. Existing opportunities are mainly open to researchers with some experience, meaning that Danish researchers are severely handicapped if compared to foreigners.

DANIDA administrators appear persistently to have had an open mind when considering research applications, seeing the usefulness of trying to reduce the risk of failed projects. The *DANIDA board* has generally - and for obvious reasons - been more hesitant in its support: political requirements mean that every Danish Krone earmarked for development assistance must be used effectively for that purpose. And the effects of research in improving development may be difficult to prove.

Other development agencies (e.g. the UN, World Bank, EEC ...) regularly announce funds for research. However, Danish awareness of this - as well as the ability to obtain them - is weak. Up to now mainly well-established research projects have been able to utilize such opportunities.

The *NGOs* usually appreciate research. It enhances their efficiency without drawing on their funds. However since NGO budgets are usually quite slim, there are few or no researchers on the staff, and requisition of research is expensive.

The worst problem is that access to external research is difficult. Recent results can mainly be obtained via periodicals and books, which are often expensive to trace and cumbersome to study.

In practice scientific 'novelties' are often only available after their 'diffusion' into textbooks and training courses. Consequently, NGOs are not to blame for duplicating prototype projects developed by others - or concentrating on projects with less 'scientific' content. Since this situation may tend to depreciate the aid extended, it is undesirable. The answer to this problem lies in the quicker dissemination of scientific results and facilitating their conversion into practical solutions. Better consultancies can help, but improved reporting should also be considered.

Within the *projects* themselves, there is a great need for research, foremost among them is that of immediate delivery [??]. Some of the cases in which trouble-shooting becomes necessary would never develop in the context of improved problem-identification and planning, though a remainder will always appear underway. A modernized project management model, emphasizing adaptive changes in activities and even in objectives, will greatly rely on a learning approach. This too, must involve research to arrive at correct solutions.

Summing up, it seems evident that the various groups involved in development assistance estimate very different quantities of research to be necessary. Interestingly, it seems that the DANIDA board has attempted a universal coverage of needs by allocating them in beautiful arithmetical proportions. Against this background, what are the prospects for a future increase in the volume of research?

Future Demands for Research Related to Development Aid

In an area where demands are not solely dependent on the require-ments of ongoing activities but also on the planning of future ones, any prediction is difficult. Selection of recipient countries and the amounts of money available are part of the political process.

In Denmark, there has been political consensus concerning the main issues regarding development aid, resulting in a fairly stable increase in the total amount of aid, now approaching the 1% of BNP target generally accepted.

The distribution of funds for multilateral (including multi-bilateral) and bilateral categories is important for Danish research. Contributions to the former tend to trigger off comparatively less Danish research than those to the latter category.

Another feature - the selection of principal aid recipient countries - is equally important. While profound changes in the multi/bilateral proportions are not likely to take place, a revised list of principal recipient countries has recently been published (February 1990, *DANIDA Nyt* No. 1). Generally, a considerable concentration on fewer countries has been decided upon for the bilateral aid.

This concentration may prove quite important for Danish research. For two reasons:
- the larger amount of funding available for the selected countries may call for a larger volume of research to be carried out.
- the need for a deeper and more coherent knowledge of the pro-blems of the host country calls for a wider range of scientific disciplines to be applied.

An evident advantage of so-called country concentration, is that a much better background analysis and problem identification is possible with resident staff and increased funding. Implementation of projects is also facilitated, since local administrative practices are better known etc. Monitoring can also be easier, especially if monitoring centers are established. By means of modern technologies, such as the use of re-mote sensing for data acquisition, can be utilized.

For research itself, extra advantages present themselves. Most notable among these is perhaps that liaison with local scientific circles can be enhanced. The advantages arising from this are obvious: local expe-rience will benefit projects and local expertise will be supported by cooperation with foreign researchers.

The single fact that concentration of strength allows projects to be much better selected can alone increase the efficiency of aid conside-rably in the short term. In a longer perspective the enhanced embedding of projects into local society from the cooperation with national research and administration could prove instrumental. For research new opportunities may occur. Those branches already active in research may in some cases see less duplicative work, and rather more innovative activities being called for. In addition, the chances are that wider circles of research will be activated from the simple fact that increased use of funding will necessitate new efficiency-orientated utilization.

Future changes in aid policy may eventually also channel more aid into the bilateral regime than was the case previously. And why not? The ghost image that too much bilateral aid money might return, irregularly, to Danish pockets is unlikely to materialize, since the control of funds is relatively easy (compared to that of multilateral means). In this connection it can be asked why the available and relevant Danish research capacity should not be more efficiently used: to create a Danish base of knowledge, to internationalize Danish research and to promote the quality of Danish development aid.

Ongoing Research Programmes

The search for plant molluscicides in the Sub-Saharan Africa by the Danish Bilharziasis Laboratory and a network of African and Danish research institutions.
Responsible: Peter Furu, Parasitologist
Danish Bilharziasis Laboratory
Jægersborg Allé 1 D
DK-2920 Charlottenlund

The Danish Bilharziasis Laboratory (DBL), a World Health Organization (WHO), Collaborating Centre for Applied Medical Malacology and Schistosomiasis Control, is a private foundation which operates as a teaching, research and service institution for water-borne parasitic diseases in man and his domestic stock, with primary emphasis on African schistosomiasis (bilharziasis), filariasis and dracunculiasis.

Schistosomiasis control is effected through a combination of various control measures, of which control of the intermediate host snails by application of molluscicides (snail killing agents) plays an important role. An alternative to the present use of expensive synthetic molluscicides is the use of molluscicides of plant origin. Advantages of the plant molluscicides are that they are grown locally and they are often more easily biodegraded than the synthetic analogues and thus more acceptable from an ecological point of view.

In order to contribute to the research on plant molluscicides, DBL has established a research network consisting of natural product chemists from the Royal Danish School of Pharmacy and the Royal Veterinary and Agricultural University, Copenhagen and biologists from the DBL. The overall aim of the field research programmes of DBL is to strengthen the research capabilities at collaborating African research institutions. Presently, projects on plant molluscicides are performed in collaboration with relevant institutions in the Sudan, Egypt, Nigeria and Zimbabwe. Some of the projects constitutes Ph.D. study programmes for African and Danish students.

A number of plants have been reported to contain large amounts of potent molluscicides. The molluscicides are found among many different types of natural products including sesquiterpenes, coumarines, flavonoids, tannins and saponins.

The present DBL-plant molluscicide projects in Africa includes agrobotanical studies on *Phytolacca dodecandra* in Zimbabwe, studies on *Balanites aegyptiaca* in the Sudan, screening of potential Egyptian plant molluscicides, and studies on the field application of *Tetrapleura tetraptera* in the control of schistosomiasis in Nigeria.

Besides the mentioned research activities, DBL has a number of other ongoing field research programmes (parasitological, malacological, epidemiological) in Ghana, Liberia, Kenya, Tanzania, Sudan, and Zimbabwe.

Monitoring and evaluation of local people's participation in renewable natural resource management projects in the Sahel: A Case Study of the Sand Dune Stabilization and Agro-Sylvo-Pastoral Development project in Mauritania".
Responsible: Søren Lund, Cand. comm.,
Ph.D. student at the International Development Studies Programme,
Roskilde University Centre,
DK-4000 Roskilde

Assessments made of renewable natural resource management projects in the Sahelian region towards the mid-eighties generally showed results to have been largely insufficient. The modest levels of local people's participation in project activities was evoked as being one of the key critical points of concern.

One of the major projects of this kind is The Sand Dune Fixation Project in Mauritania. Implemented by the FAO, the project was launched 1983 on a joint UNSO/DANIDA, UNDP, WFP, and UNCDF funding. In the course of its first phase, 1983-86, the Sand Dune Fixation Project achieved positive technical results and developed a contractual model of collaboration with the local communities.

Based on the experiences gained during the first phase, this research project was originally intended as a contribution towards the elaboration of an M&E model concerning the local people's participation in project activities during the second phase of what is now called the Sand Dune Stabilization and Agro-Sylvo-Pastoral Development Project in Mauritania.

As the actual work has proceeded, new theoretical insight as well as practical considerations in connection with data-collection and working conditions in Mauritania have made a shift in the focus of the research necessary. More emphasis will be placed on the understanding of the local people's participation as an expression of the relationship between local actors and social structure.

Publication: Peter L. Wright, Jan Broekhuyse, Fred R. Weber, and Søren Lund: *Ecology and Rural Development in Sub-Saharan Africa: Selected Case Studies.* Club du Sahel, OECD. Sahel D(89)327. Paris, 1989.

Practical work: 1) Participating as an external consultant for the Danish Red Cross in the planning and monitoring of environmental education and training projects for youth and children in Senegal, Burkina Faso, and Sudan. These projects are conceived and implemented as "participatory projects" with local institutions playing the major role. 2) Consultancy work appraising/monitoring Red Cross in Senegal; appraising UNSO integrated rural development project in Burkina Faso.

Mapping and monitoring of soil degradation indicators by means of digital satellite image interpretation.
Responsible: Pia Frederiksen,
Institute of geography, socioeconomic analysis and computer science,
Roskilde University Center,
DK-4000 Roskilde

The repetitive coverage of the surface of the earth by satellites carrying multispectral scanners, offers a potential for mapping and monitoring of various surface features.

In recent years, attention has been drawn towards the possibility of using satellite images for investigations of the ecological degradation occurring in various arid and semiarid tropical lands. Among others, the following aspects have been emphasized:
- the lack of base maps in large parts of tropical and subtropical rangelands combining with the relatively low costs of spatial coverage from satellite images, to make these attractive for small-scale mapping of natural resources.
- the potentials of using continuous recording for monitoring purposes. The current discussion of land degradation and desertification - the climatic impact and the possible reversal of these processes - illustrate, that actual knowledge of the ecological interaction and development in these areas is still frag-mented, and more information of processes is needed.

Project area: The present work is based in a semi-arid bushland situated on the margins of the agriculturally utilized land in Kenya. Kitui District is inhabited by the Kamba tribe - originally a nomads, some of which, coming from the south, settled in the highlands and on the inselbergs of Kitui. Here the homesteads and gardens were established, while the lowlands, where rainfall is less and water resources generally scarce, was used as wet season grazing for the cattle. Increased pressure on the land has lead to almost total deforestation and cultivation of the highlands and increased permanent settlement and cultivation of the lowlands, disturbing the former complemen-tarity of the two ecological landtypes. Cultivation without soil conservation and overgrazing due to unimproved grazing management combined with de-creasing grazing area has lead to accelerated ecological degradation - e.g. bush encroachment, almost total eradication of perennial grasses and soil erosion.

Project components: The aim of the project is to contribute to the development of methodologies for using satellite images to map information of soil and vegetation related factors in rangelands. Comparison of images from different years yield information on changes in these factors, which may be interpreted according to soil degradation. Owing to the close relationship between vegetation cover and soil erosion *estimation of vegetation cover in a rangeland scene* has been a major component investigated. Two aspects have been emphasized here:

- the necessity of estimating defoliated vegetation and litter as well as green vegetation, since the bushland often appears as a mixture.
- compensation for soil background influence on the vegetation spectral signature.

Visible changes in the soil surface due to the removal of organic matter and finer particles through the soil erosion process have lead to investigation of *spectral signatures of soil surfaces influenced by soil erosion.* Two periods of fieldwork have now created the true ground basis for the analysis. Radiometer measurements have been conducted in the field and in the laboratory, and soil samples are being analyzed for various factors possibly influencing the spectral responds of soil. Images from 1979, 1987 and 1989 will be analyzed and the true ground picture from 1979 will be reconstructed from aerial photos taken in 1980.

Funding: The project is part of a Ph.D. study funded by the Danish Council for Development Research.

Publication: Pia Frederiksen: Mapping and monitoring of soil degradation indicators by digital satellite image interpretation, working paper 1: model building. Working Paper no. 81, Institute of geography, socioeconomic analysis and computer science, RUC 1989.

A satellite based analytic framework for agricultural and environmental monitoring in Sahel - case study Oudalan, Burkina Faso.
Responsible: Kjeld Rasmussen and Anette Reenberg,
Institute of Geography,
University of Copenhagen,
Øster Voldgade 10,
DK-1350 Copenhagen K.

The project aims at analyzing the development and structure of agricultural systems in semi-arid environments. A satisfactory analytical framework for these agricultural systems must include environmental as well as socioeconomic parameters. Such data are very often not easy obtainable, and potential data-sources has to be evaluated with respect to accuracy and "costs". An analytical frame-work for a satisfactory detailed mapping and monitoring of semi-arid agricultural systems has been presented in Rasmussen and Reenberg, 1989 (AAU Reports 19, Botanical Institute, University of Aarhus).

The component dealt with during fieldwork in October-November 1989 is SPOT-satellite-image based mapping of land-use and particularly millet-acreage. The work includes ground-truth mapping in the northern part of Oadalan Province and village interviews in 10 selected villages in the region - the latter focusing on an evaluation of the fluctuations in the cultivated area and bottlenecks in the agricultural system.

Assessment of millet yield in Northern Burkina Faso on the basis of satellite images.
Responsible: Michael Schultz Rasmussen,
Institute of Geography,
University of Copenhagen,
Øster Voldgade 10,
DK-1340 Copenhagen K

Monitoring of vegetation and crops in the Sahel requires use of a combination of high resolution satellite images, which are acquired at long time intervals, and medium-to-low resolution images, which are acquired at intervals of hours or days.

In the case of millet-fields in northern Burkina Faso, which has served as a test area for this project, SPOT HRV data, with a spatial resolution of 20 x 20 m., have been used to locate and assess the size of the cultivated areas, which may then be identified in the NOAA AVHRR data, which have a spatial resolution of 1 x 1 km.

For those areas in the NOAA images which have been identified as fields, the development of "greenness" of the crop through the growing season may be monitored, using the concept of the "normalized difference vegetation index" (NDVI). By integrating the NDVI over the last part of the growing season an estimation of crop yield may be made.

Since the cropped area may be estimated from the SPOT data, as studied in detail by another project reported here, the total grain production within the region may be assessed. This may provide an input into "early warning" systems.

The method has been tested for the growing season of 1988, but in order to fully verify the potential of the methodology, more studies will be required, in particular aiming at investigating the effect of interannual variations in crop development and atmospheric conditions.

The work will be continued, and in particular the use of satellite-derived evapotranspiration data as an estimator of vegetation/crop productivity will be studied. It is believed that this may prove more sensitive than methods based on greenness monitoring in areas with sparse vegetation, such as the

Sahel. Emphasis will probably shift towards monitoring of natural vegetation in addition to crops. By introducing a digital terrain model also mountainous areas will be attempted monitored.

It is planned that the further work will be carried out within the framework of a Ph.D.-study.

Conservation and development of the Air Mountains and the Tenere Desert, Niger.
Responsible: Kim Karstensen
WWF Denmark,
Ryesgade 3 F,
DK-2200 Copenhagen N

In January 1988 the Air-Tenera area was declared a National Nature Reserve by the Government of Niger. The area comprises some 77,000 km2 covering two distinct environments and the hyper-arid Tenere Desert.

The area is the home of about 4,500 people, of which some 2,500 are sedentary while the rest are nomadic or semi-nomadic. The population fluctuates, however, as a function of the availability of grazing resources.

Because of the geographical position between Sahara and Sahel, and the oasis-like character of the Air Mountains, the area is very diverse in a biological sense, compared to its more homogenous surroundings. In spite of serious overutilization of the natural resources during the droughts of the 70s and the 80s, the area ha s retained most of its biological values, and today it must be said to be quite unique in the region. During the last few years trees and perennial grasses have clearly been recovering, due to three consecutive years of reasonably good rains.

Together with IUCN - the World Conservation Union - WWF has been working in the Air-Tenera area since 1979. Our initial interest was the protection of the fauna in the area (addax, ostriches, barbary sheep, gazelles), but gradually our approach has been broadened. The project now focuses on the interface between the physical (soils, water, flora, fauna) and human environments, with a view to promote the sustainable utili of the resources and the satisfaction of the needs of the human population.

The project is presently reaching the end of a phase, where it has been funded primarily by the Swiss development agency, DDA. This phase has allowed the project to develop its human oriented activities significantly. Among other things the project has developed techniques for woodless construction, for relatively cheap well protection and deepening, for garden protection, and it has gained experience with rehabilitation of pastures and with reafforestation in the area.

We hope to be able to continue the project in a new 4-year phase, which should allow us to consolidate the results already achieved, and to establish a coherent management plan for the area.

The management plan should be based on three axes of activities:
- the establishment of a research based monitoring system, covering both the natural resources (by means of satellite imagery, aerial surveys, and more traditional ecological methods) and socio-economic questions (herd sizes, strategies of resource utilization, diversification of incomes, etc.).
- experiments with rehabilitation of resources (like for instance introduction of rational grazing systems or priority consumption and annual grasses in good rainfall years) or introduction of improved techniques in resource utili (improved stoves, new types of mortars, etc.).
- consultations with the local population, and with authorities.

We hope the project will provide both us and the wider donor and research environment with valuable experience on ways to combine conservation interests with the satisfaction of the immediate needs of the human populations, and with their demands for continued access to resources, security, economic progress or the like.

Livestock Development Study in the Sahel
Responsible: Ole Olsen, Agronomist and Christian Lund, Socio-economist and Geographer,
Danagro Adviser A/S.
Granskoven 8,
DK-2600 Glostrup

The "Livestock Development Study in the Sahel to assess the Potential Role of Livestock in Integrated Farming Systems" in Senegal, Mali, Burkina Faso and Niger that Carl Bro International was undertaking in 1989, was taken over by Danagro Advisers (a new firm in the Carl Bro Group) for completion i 1989. The finalization concentrated on the socio economic aspects of the study and produced an analysis where the comprehensive assessment of social groups' strategies and constraints was the key issue.

Much attention was given to the social structure of the studied areas in order ti identify patterns of social development, and much attention was given to see the strategies pursued by one group of farmers as affecting the range of possibilities of other groups of farmers. Thus the enablement of one often proved to be the constraint of the other.

The samples showed great social diversities. The diversities were, however, *not infinitely numerous or as accidental* as they may have seemed by first appearance, but ordered after a number of principles.

It has been possible to identify social groups in each country sample with approximately the same physical, economic and social constraints, and by examining the socio-cultural relations prevailing in the rural communities, the understanding of the principles of social differentiation has improved and patterns has emerged.

The *Senegalese case* thus demonstrated that homogenous social systems

124

simultaneously enabled wealthy farmers to thrive through their dependents' need for a social "safety net" and secured the poorest farmers from total pauperization. Compared with villages of a heterogeneous social appearance the socially homogeneous villages displayed a significant higher economic dispersion between the households as well as a lower limit as to how poor the poorest farmers has become during the recent general economic crisis. In the wealthier villages the poor farmers had managed to maintain their animal traction, whereas in the poor villages there seemed to be no lower poverty limit.

To this picture *the Malien case* added a facet. Whereas the wealthy farmers' social relations with poor farmers helped the poor farmers to maintain ox traction in the wealthy villages, the relations between the two groups in poor villages did not enable the poor farmers to maintain ox traction. On the other hand the wealthiest households' maintenance of ox traction in these poor villages were possible through the relations between them and their poorer dependents.

The *Nigerian Case* gave an example of how social coherence alone did not carry much weight when the general economic conditions did not leave margin for surplus extraction at all. Very few farmers could pay anything in return for labor - nor even social safety - and therefore it appeared that very few could in any way profit from the social coherence.

Strong social coherence was thus not a guarantee for animal traction or any form of integration between livestock and agriculture if the economic conditions are sufficiently adverse, but, on the other hand, crop prices did not necessarily have to be so high that they alone could sustain animal traction in a household if strong social coherence was provided from the community. This indicated that the social coherence provided a cushion and a considerable time lag against the impact of economic adversities, but that the *general trend could be a de-mechanization of the farming systems if no interventions are made.*

The *Burkina Faso* case provided some fuel for a hypothesis concerning social relations among the very prosperous farmers. The wealthy farmers generally profited largely from poorer dependents' work on their fields like in the case of Senegal and Mali, but the high profits from cotton had enabled some of the wealthiest farmers to use pesticides and tractors, thus, reducing their dependence on kin labor and de facto depleting the social relations of their economic and productive contents. Thus, above a certain economic threshold, socio-cultural relations may decrease in importance, and the societies may consist more of individual farmers than be social systems of inter-dependence.

The range of variation calls for tailor made and flexible interventions, and the detailed design should of course take place in situ. However, the data made it possible to identify social groups and to put forward certain considerations concerning *targeting* i.e. who would need which interventions in order to secure mixed farming practices, esp. animal traction.

The data showed that a smaller number of *wealthy farmers* would not need any assistance from donor agencies to maintain and improve their situation.

They would, on the other hand, probably object to interventions alleviating the burdens of the poorer segment of farmer since the dependence of the latter to a degree is securing the wealth of the former. Poverty alleviation could then be seen as implemented at the detriment of wealthier farmers.

Another segment of farmers, the *medium sized farmers*, generally had animal traction but were in an exposed position and only maintained their animal traction due to good social relations. The nigerian example showed, however, that the socio-cultural insurance relations might only be a limited respite which, when exhausted will give place pauperization. To this group of farmers the *institutional marketing and savings strengthening* efforts would be of the out most importance to assure the *continued maintenance* of animal traction by this social group.

The third and *poorest group of farmers* without animal traction is rather difficult to reach. Projects targeted to the poor are notorious for missing their targets and being seized by the wealthy social segments.

The findings of the general conclusions indicated that there was a significant difference in obtaining and maintaining animal traction. The kinship based insurance system often enabled poorer farmers to maintain animal traction by helping to prevent distress sales, but in none of the studied households had the kinship relations enabled poor farmers to buy traction animals. This would advocate that the poor farmers were *basically assisted in the purchase of traction animals*. It should be borne in mind, however, that an important reason for the significant difference in obtaining and maintaining animal traction to a particular household were its social network, and that the farmers in this group to some extent were characteri by the lack of it. If, however, other measures as the institutional strengthening of marketing and savings facilities were taken, they probably could provide some degree of substitution for the kinship based "safety net".

Other activities in the Sahel carried out by members of the Carl Bro Group in 1989 include:

Water supply schemes in Burkina Faso. Identify and appraise water supply projects for 9 towns and 3 urban areas. Feasibility study, design and supervision of semi-urban water supply. .

Ethiopia fuel wood. Baseline studies for the establishment of fuelwood plantations and for integrated rural development of the Blue Nile ecological zone.

Djibouti-Somalia road. Revision of the tender documents and supervision.

Somalia four towns power rehabilitation project.

Somalia Wind Power. Evaluation mission.

Sudan, Gezira. Construction of dairies; management and supervision.

Sudan, Gezira. Supervision of power rehabilitation.

Sudan water yards. Rehabilitation of 150 water yards.

Sudan el Obeid en Nahud Road. Feasibility study of construction of 200 km road.

126

Sudan - restocking of the Gum Belt. Preparation of project.
Mali, Macina-Markala road. Feasibility study and supervision of construction of 100 km road.
Mali Livestock. Baseline Study of the livestock sector.

Vegetation changes
Responsible: Christina Skarpe,
Institut of Ecological Botany
Uppsala University,
S-75122 Uppsala

Indications of changes in the vegetation area in the arid and semi-arid zones of Africa can be obtained from (sub)fossil and archaeological evidence, oral tradition, written records, living memory and direct monitoring and research. Reasons for change can be geological or evolutionary processes, climatic change, human land use or interaction between all. As the timespan considered gets shorter and more recent, mans direct and indirect contribution increases.

Prehistoric rock paintings in Sahara illustrate some aspects of a tradition from lush savanna with elephants and buffalos to arid conditions with ostriches and camel riders during some 10,000 years of climatic change and human impact. The development is presently mimicked in the corresponding ecological zones of Southern Africa in less than 1/10 of the time, and with little evidence of climatic change.

For resource development and land use strategies, the main interest may be on changes during the last decades or, at most, century. In that time perspective, in most cases, human impact is considered the prime agent. (there may be too little data to tell the nature of the decrease in rainfall in the Sahel since the mid 1960's). In a short time perspective, it is important to know the extent and rate of the ongoing vegetation changes. In a longer time perspective, the emphasis should be put on the mechanisms of the changes including the complete interactions between land use strategies and ecological determinants.

Tropical areas with seasonal rainfall and annual means of ca. 2000 to 200 mm, usually support savanna and savanna like vegetation. Four main determinants for these vegetation are soil water, soil nutrients, fire and herbivore. Mans activities directly affect frequency and times of fires as well as kind and degree of herbivore. This indirectly influences the availability of and competition for water and nutrients. Man also cuts trees and cultivate fields, thus further affecting the system.

Vegetation changes recorded from arid and semi-arid land often include the formation of entirely new vegetation types, and are not always easily reversible. There are theories of some grazing ecosystems under arid or semi-arid conditions having more than one equilibrium point or limit cycle, or even being totally event driven. The model with multiple equilibrium points explains the rather stable nature of the degraded vegetation types. A return to the original dynamic equilibrium may be possible, but is complicated by changes in soil properties, microclimate, availability of diaspors etc, accompanying the change in vegetation.

A change in physiognomy towards a more shrubby vegetation at the expense both of open grassy areas and of trees, is reported from many dry rangelands, both tropical and temperate. Evidence range from comparisons with up to a hundred year old descriptions of land and vegetation to monitoring and experiments during a few years. Changes may be a diffuse increase in the woody vegetation or a massive invasion of shrubs with up to 100% canopy cover. The change may not mean a decrease in phytomass or production per se, but the new vegetation has low value for most human use. The change is usually attributed to human disturbance, mainly overgrazing by livestock, but also to changes in fire regime, decrease in indigenous browsers, shifting cultivation or cutting of trees. Weather conditions and climatic changes have also been held responsible, alone or in interaction with other disturbances. The driving mechanism may include a change in amount and distribution, in time and space, of soil water, in combination with a shift in competitive advantage from grasses to the shrubs also leading to increased competition within the woody vegetation.

There are also examples of savanna-like vegetation changing into completely open herbaceous vegetation, presumably following overutilization.

Changes within the herbaceous vegetation includes shifts in species and life form competition. The direction of the change probably depends mainly on the kind of disturbance, making different attributes vital for survival. A certain land use, usually implies many different factors of disturbance, e.g., pastoralism may include, apart from the complex grazing factor itself, burning, cutting the wood, destruction of indigenous herbivores etc. (Over-)grazing causes a continuous or step wise change from palatable to unpalatable species, usually including a decrease in perennial grasses and an increase in perennial forbs or dwarf shrubs, meaning a general decrease in grazing value. Changes from perennial to annual grasses and/or forbs have also been reported or assumed. The effect of fires depend much on time of the year and on environmental conditions. Except in the driest areas, frequent fires causes a decrease in the woody component. Abandoned field successions are often started by annual weeds or by shrubs.

The change in fertile land into desert, devoid of (almost) all vegetation, is the main concern for the study of "desertification". The extent, speed and significance of this "spreading of deserts" is much discussed. There is evidence that the generally accepted figures are exaggerated, and, perhaps, do not clearly distinguish between temporary fluctuations, depending mainly on variations in rainfall, and long term trends. Severe vegetation and land degradation may take place at population centers also far from the edges of existing deserts. It has been claimed, that the most significant and wide spread problem is not the "desertification" proper, but the degradation of vegetation to more or less persistent, from human stand point low productive, types.

Bush-fire monitoring in the Ferlo area of Senegal
Responsible: Sindre Langaas,
Institute of Geography,
University of Copenhagen
Øster Voldgade 10,
DK-1340 Copenhagen K, and
Centre de Suivi Ecologique, Dakar.

Bush-fires constitute a major environmental problem in the Sahelian part of Senegal, both because it causes loss of fodder resources, and because it may affect the tree cover negatively. Methods for monitoring of fire-affected areas as well as of ongoing bush-fires are therefore of great interest. NOAA AVHRR satellite images provide a convenient basis for such monitoring, since images are acquired almost every day and night. The present project aims at developing and testing methodologies for such monitoring. While fire-affected areas may be seen on images acquired during the day, ongoing fires may best be identified on images acquired at night. The middle- and thermal infrared channels of NOAA AVHRR allow calculation of fire temperature, which has considerable ecological significance, and the extent of the fire with subpixel accuracy. NOAA-based bush-fire monitoring will, when the results been verified through field work, be included as an element in CSE's operational activities.

Cooperation between Centre de Suivi Ecologique, Dakar, and Institute of Geography, University of Copenhagen, on satellite-based agricultural and environmental monitoring
Responsible: Henrik Steen Andersen, Kjeld Rasmussen, and Inge Sandholt.

Since 1987, CSE and GI have cooperated on development of methodologies satellite-based monitoring of vegetation, bush-fires and agrometeorological parameters of relevance to CSE's operational activities, which have the primary objective to monitor the grazing system of the Ferlo area of northern Senegal. NOAA AVHRR images play a major role as a basis for biomassestimation and assessment of areas affected by bush-fires. Meteosat images are used as a basis for rainfall-monitoring. It is planned that CSE will move towards applications of high-resolution satellite data for land use mapping in millet- and groundnut-cultivation areas.

Thus GI is developing and testing methodologies and software for applications of a variety of types of satellite images for agricultural and environmental monitoring. All software has been assembled in a publicly available software package, CHIPS, which forms the basis of CSE's activities, and which is being used by several other institutions in developing countries.

Refugee Woman in Somalia
Responsible: Ingeborg Kragegaard Higashidani
Department of Ethnography and Social Anthropology,
Aarhus University
DK-8270 Højbjerg

I have carried out fieldwork among refugees at various refugeecamps and settlement schemes in Somalia in the period from december 1985 till november 1986. At the moment I am writing my final field-report, titled "Refugee Woman in Somalia - a study of the social and cultural changes in refugee camps with special reference to the position and role of Woman". The main point of the report is to uncover the impact of the ecological and political development on the social organization in general, and more specifically on the situation of woman.

Swidden Agriculture and Social Structure
Responsible: Søren Leth Nissen
Department of Ethnography and Social Anthropology,
Aarhus University
DK-8270 Højbjerg

I am at the moment finalizing my M.A. thesis about various explanations of the impact of ecological surroundings on the shaping of social structure in different societies, mainly societies practicing swidden agriculture. I have at the same time worked with various ways of how to counteract the ongoing landdegradation in general as well as done some ethno-botanical survey. My main field experience is from Latin America.

Development Strategies in Burkina Faso
Responsible: Inge Schou
Department of Ethnography and Social Anthropology,
Aarhus University
DK-8270 Højbjerg

I have done field research in Burkina Faso in 1987 and visited the area several times since. The main topic of my field work was to describe the social consequences of a small scale development project, mainly aimed at

130

improving the economic abilities for woman. At the moment I am writing my final Ph. D. thesis on the subject of various development strategies.

Landed Property in Algeria
Responsible: Jo Falk Nielsen
Department of Ethnography and Social Anthropology,
Aarhus University
DK-8270 Højbjerg

For the time being I am preparing my M.A. thesis. The subject concerns the changes in landed property in Algeria and its importance for the development of the algerian society, thereby touching upon the relationship between grain-growing and cattle-breeding societies, the problem of monoculture, soil-exhaustion and -erosion and the rural exodus.

131

Participants

Alstrup, Vagn
Inst. Ecological Botany,
Østre Farigmagsgade 2 D,
1353 Copenhagen K.

Benjaminsen, Tor Arve
Geographical Institute,
Oslo University,
Postbox 1042, Blindern,
N-0316 Oslo

Bensoala, Tine
Institute of International Development Studies,
RUC,
DK-4000 Roskilde

Bovin, Mette
Piskesmældet 8, 2 tv.,
DK-3000 Helsingør

Brauer, Ole
Folkekirkens Nødhjælp,
Copenhagen

Brimer, Leon
Inst. Pharmacology and toxicology,
Bülowsvej 13,
DK-1870 Frederiksberg C.

Buck, Jette
DANIDA,
Asiatisk Plads 2,
DK-1448 Copenhagen K

Christiansen, Sofus
Inst. Geography,
Østre Voldgade 10,
DK-1350 Copenhagen K.

Degnbol, Tove
Cowi-Consult,
Parallelvej 15,
DK-2800 Lyngby

Dittmer, Grethe
DANIDA,
Asiatisk Plads 2,
DK-1448 Copenhagen K.

Frederiksen, Dan
Hedeselskabet,
Postbox 110,
DK-8800 Viborg

Frederiksen, Peter
Inst. Geography, Socio economic
analysis and computer science,
RUC,
DK-4000 Roskilde.

Frederiksen, Pia
Inst. Geography, Socio economic
analysis and computer science,
RUC,
DK-4000 Roskilde.

Furu, Peter
Dansk Bilharziose Laboratorium,
Jægersborg Allé 1 D,
DK-2920 Charlottenlund.

Geertsen, Jette
I. Krüger A/S,
Gladsaxevej 363,
DK-2860 Søborg.

Genefke, Hans
DANIDA,
Asiatisk Plads 2,
DK-1448 Copenhagen K..

Graudal, Lars
Danida Forest Seed Center,
Krogerupvej 3 A,
DK-3050 Humlebæk.

Higashidani, Ingeborg Kragegaard
Dept. Ethnography and Social An-
thropology,
DK-8270 Højbjerg.

Jensen, Dorrit Skov
Institute of International Develop-
ment Studies,
RUC,
DK-4000 Roskilde.

Koed, Jens
Botanisk Institut,
Nordlandsvej 68,
DK-8240 Risskov.

Kvist, Lars Peter
UniConsult International,
Science Park Aarhus
Gustav Wiedsvej 10,
DK-8000 Århus C.

Langaas, Sindre
Inst. Geography,
Østre Voldgade 10,
DK-1350 Copenhagen K..

Lawesson, Jonas Erik
Botanisk Institut,
Nordlandsvej 68,
DK-8240 Risskov.

Leth-Nissen, Søren
Dept. Ethnography and Social An-
thropology,
DK-8270 Højbjerg

Lund, Christian
Danagro Adviser,
Granskoven 8,
DK-2600 Glostrup.

Lund, Søren
Institute of International Develop-
ment Studies,
RUC,
Postbox 260,
DK-4000 Roskilde.

Lundberg, Hans Jørgen
DANIDA,
Asiatisk Plads 2,
DK-1448 Copenhagen K..

Manger, Leif
Senter for Udviklingsstudier,
NORAD,
Bergen University,
Strømgaten 54,
N-5007 Bergen

Mostrup, Søren
Danida Forest Seed Center,
Krogerupvej 3 A,
DK-3050 Humlebæk.

Nielsen, Holger Koch
I. Krüger A/S,
Gladsaxevej 363,
DK-2860 Søborg.

Nielsen, Ivan
Botanisk Institut,
Nordlandsvej 68,
DK-8240 Risskov.

Nielsen, Jo Falk
Dept. Ethnography and Social Anthropology,
DK-8270 Højbjerg.

Nyholm, Klaus
DANIDA,
Asiatisk Plads 2,
DK-1448 Copenhagen K..

Ollson, Lennart
Dept. Physical Geography,
Lund University,
Sölvegatan 13,
S-223 62 Lund

Olsen, Ole
Danagro Adviser,
Granskoven 8,
DK-2600 Glostrup.

Poulsen, Ebbe
Dept. Ethnography and Social Anthropology,
DK-8270 Højbjerg.

Rasmussen, Kjeld
Inst. Geography,
Østre Voldgade 10,
DK-1350 Copenhagen K..

Rasmussen, Michael S.
Inst. Geography,
Østre Voldgade 10,
DK-1350 Copenhagen K..

Reenberg, Anette
Inst. Geography,
Østre Voldgade 10,
DK-1350 Copenhagen K..

Schou, Inge
Dept. Ethnography and Social Anthropology,
DK-8270 Højbjerg.

Schønnemann, J.
CowiConsult,
Parallelvej 15,
DK-2800 Lyngby.

Seddon, David
International Institute of Environment and Development (IIED),
3 Endsleigh Street,
London WC1H 0DD

Skarpe, Christina
Inst. Ecol. Botany, Växtbio,
Uppsala University,
S-75122 Uppsala

Speirs, Mike
Økonomisk Institut,
KVL,
Thorvaldsensgade 40
DK-1871 Frederiksberg C..

Styczen, Merete
Dansk Hydraulisk Institut,
Agern Allé 5,
DK-2970 Hørsholm.

Toulmin, Camilla
International Institute of Environment and Development (IIED),
3 Endsleigh Street,
London WC1H 0DD

Tybirk, Knud
Botanisk Institut,
Nordlandsvej 68,
DK-8240 Risskov.

Ørum, Thorkild
I. Krüger A/S,
Gladsaxevej 363,
DK-2860 Søborg.